THE C

NODE
BOOK

About the Author

Kevin Burk (California) holds a Level IV NCGR certification in astrological counseling and has practiced astrology in San Diego since 1993. His first book, *Astrology: Understanding the Birth Chart*, teaches classical astrology techniques, and is used as a basic astrology textbook at Kepler College. His website has won several Internet awards and has hosted more than 1.5 million visitors.

To Write to the Author

If you wish to contact the author or would like more information about this book, please write to the author in care of Llewellyn Worldwide and we will forward your request. Both the author and publisher appreciate hearing from you and learning of your enjoyment of this book and how it has helped you. Llewellyn Worldwide cannot guarantee that every letter written to the author can be answered, but all will be forwarded. Please write to:

Kevin Burk
‰ Llewellyn Worldwide
P.O. Box 64383, Dept. 0-7387-0352-4
St. Paul, MN 55164-0383, U.S.A.

Please enclose a self-addressed stamped envelope for reply,
or $1.00 to cover costs. If outside U.S.A., enclose
international postal reply coupon.

Many of Llewellyn's authors have websites with additional information and resources. For more information, please visit our website at http://www.llewellyn.com

THE COMPLETE

NODE BOOK

UNDERSTANDING YOUR LIFE'S PURPOSE

KEVIN BURK

2003
Llewellyn Publications
St. Paul, Minnesota 55164-0383, U.S.A.

First Edition
First Printing, 2003

Book design by Donna Burch
Cover image of antique celestial map © 2002 by Visual Language
Cover design by Kevin R. Brown
Editing by Andrea Neff
Interior illustration on page 5 by the Llewellyn art department

Library of Congress Cataloging-in-Publication Data

Burk, Kevin, 1967–
    The complete node book : understanding your life's purpose / Kevin Burk.—1st ed.
        p.   cm.
    ISBN 0-7387-0352-4
    1. Astrology. 2. Moon—Miscellanea. I. Title: Node book. II. Title.

BF1723.B87 2003
133.5'32—dc21   2003044670

Llewellyn Publications
A Division of Llewellyn Worldwide, Ltd.
P.O. Box 64383, Dept. 0-7387-0352-4
St. Paul, MN 55164-0383, U.S.A.
www.llewellyn.com

Printed in the United States of America

Other Books by Kevin Burk

Astrology: Understanding the Birth Chart
(Llewellyn Publications, 2001)

Acknowledgments

This book, or at least the original idea to write this book, began in 1995, when I took my first professional astrology class with Terry Lamb. Terry is the person who suggested that I write something on the Moon's nodes, and she commissioned an article from me ("A New Look at the Moon's Nodes") for her now-defunct newsletter, "The Quarterly Eclectic." Terry's encouragement, along with the support of fellow students Adam Pratter, Cindy Yurkovitch, and Deane Driscoll, helped me expand the article to something more substantial, and Terry soon invited me to lecture on the Moon's nodes at a meeting of the San Diego chapter of the National Council for Geocosmic Research (NCGR).

Several years later, I was again invited to lecture on the Moon's nodes, this time by the San Diego Astrological Society (SDAS), and this invitation prompted me to spend time revising and refining my original approach and the original lecture given to NCGR. My thanks to Jim Shawvan, Pam Ciampi, and Debora Parker at SDAS. Meanwhile, I wrote another book on classical birth chart interpretation (*Astrology: Understanding the Birth Chart*), which included a revised version of my original article on the Moon's nodes.

It's been a rather long road to complete this book. My thanks to everyone who has encouraged and supported me in the writing of it.

Contents

A Note About This Book

My approach as an astrology teacher is to help my students understand the vocabulary and the grammar of the language of astrology, so that they can learn to speak it themselves and create their own individual interpretations. I strongly urge my students to avoid relying on astrological "cookbooks" that interpret individual aspects or planetary positions for them. Needless to say, I was somewhat ambivalent about writing my own "cookbook" on the Moon's nodes. My biggest beef with most astrological cookbooks is that they can never give even a reasonably complete picture—they do not factor in the combined influence of sign and house placement when doling out their interpretations. Even worse, every cookbook on the Moon's nodes that I have encountered equates the signs with the houses, and states that, for example, having the North Node in Aries is the same thing as having the North Node in the 1st house. My primary objective in writing this "cookbook" was to address this extremely damaging bit of misinformation, and to illustrate how the sign placement and the house placements must be combined to create an interpretation.

While this book does interpret all 144 possible placements of the Moon's nodes, please remember that these are my interpretations. They are not necessarily your interpretations, nor should they be. And they are certainly not the only interpretations. I have tried to provide the necessary building blocks that will enable you to understand how to create your own individual interpretations of the Moon's nodes. In chapter 1, you will learn how to understand the nodes and how to approach them in a general way. In chapter 2, you will learn how to look at the houses. For each sign axis in the subsequent chapters, I try to paint a picture of the core concerns, and how the energy and lessons of the two signs interact with each other. These are the building blocks that I used to create all of my interpretations, and I hope that you will be able to use them to create your own interpretations as well.

1

INTRODUCING THE
MOON'S NODES

The Moon's nodes are probably the most misunderstood points in astrology. Although few astrologers would dispute their importance in the chart, equally few astrologers could offer a modern, supportive interpretation of the nodes, what they represent, and why. Traditional astrology takes a rather consistent view of the nodes. The North Node (*Caput Draconis*, which means the "Dragon's Head") was given a similar quality to the traditional interpretations of Venus and Jupiter, and was associated with all of the good things that we could possibly experience in our lifetime, including success, advancement, increase, and personal fulfillment. The South Node (*Cauda Draconis*, which means the "Dragon's Tail") was of the same nature as Mars and Saturn, and was associated with huge heaping amounts of what one would expect to come out of the tail end of a dragon. The traditional interpretations of the Moon's nodes pretty much boil down to "North Node *good;* South Node *bad.*"

As modern astrology moved away from the extremely fatalistic and often very negative traditional style of interpreting the planets, the nodes too received a facelift of sorts to make working with them more empowering. The North Node became the processes and experiences that we must strive for in order to work with our karma and to grow in this lifetime. And the South Node got promoted from evil incarnate to the point in the chart where we're most likely to take the easy way out and to rely on habit. The South Node is also related to the karma that we're working off in this lifetime. In other words, "North Node *good;* South Node *bad.*"

One of the difficulties in coming up with a truly comprehensive understanding of what the nodes represent in the natal chart is that the Moon's nodes are the only points in the chart that do not have an associated Western mythology. The associations of the dragon with the Moon's nodes comes from Hindu mythology, and while working with this myth is a great help in understanding the nodes as they are used in Eastern astrology, the Hindu myth doesn't help us come up with a Western, humanistic understanding of the nodes.

If we want to understand the energy of one of the zodiac signs, for example, we can simply break it down into its component parts. Is it cardinal, fixed, or mutable? Is it earth, air, fire, or water? When we combine our understandings of the elements and the modalities, we can easily come up with an accurate understanding of the energy of each of the signs. In a similar vein, by examining the things that we can observe about the Moon's nodes and interpreting them individually, we can come up with a much more structured, comprehensive, and, above all, supportive approach to interpreting the Moon's nodes.

Let's review what we know about the nodes so far. The nodes are mathematical points that represent where the orbit of the Moon around the Earth crosses the ecliptic (which is the apparent path of the Sun around the Earth). The North Node is the point where the Moon's orbit rises above the ecliptic, and the South Node is the point where the Moon's orbit falls below the ecliptic. The North Node and the South Node are always exactly opposite each other in the chart.

This information about the nodes may not appear to be as helpful to interpreting them as the elements and modalities are to interpreting the signs; however, just this simple physical description of the Moon's nodes can help us gain a more complete understanding of what they represent in astrology.

The Nodes Are Mathematical Points— They Are Not Physical Bodies

The nodes are mathematical points; they are not physical bodies. What this means is that the nodes do not emanate light. The nodes can *receive* aspects from the planets, but they cannot directly influence how a physical body expresses itself.

This also means that the nodes do not filter the energy of the signs. With the nodes, as with the angles, we have a pure expression of the energy and symbolism of the signs. Our experience of the energy of Aries, for example, is very different when we are experiencing Venus in Aries than when we are experiencing Mars in Aries. The personalities of the planets color the expression of the signs that they visit. The nodes, however, do not change how the signs are expressed or experienced.

The Nodes Are Related to the Moon, the Sun, and the Ecliptic

The nodes are most closely related to the symbolism and processes of the Moon because they represent points on the Moon's orbit around the Earth. But the nodes are also related to the Sun because of their relationship to the ecliptic. In other words, the nodes are the points where aspects of the lunar and solar principles connect.

Let's look at the Moon first, and get a feel for what the Moon brings to this process. The Moon reflects the light of the Sun; it is passive, receptive, and feminine. The Moon responds, and produces emotions and feelings. The Moon is the container of our experience, providing form and location for the Sun's expression. The Moon relates to our conditioning, habits, vices, and learned responses; in other words, the Moon is our memory. The Moon is not just our memory of this lifetime; it is our soul's memory, and it is what our soul wants us to remember from other lifetimes. The Moon is our unconscious and our subconscious.

The Sun, on the other hand, is our conscious, active life force. The Sun is our will, our power, and our sense of purpose. The Sun is the heart of our existence; it is the motivation for our life this time around. The Sun is how we are seen, how we shine, and how we express and project ourselves. The Sun is how we want to be a hero in our life; it is how we want to become an individual.

The ecliptic, the apparent orbit of the Sun around the Earth, describes the path that our journey will follow in this lifetime. When we look at a natal chart, what we're actually looking at is a two-dimensional representation of the positions of the planets as viewed from the Earth, flattened to the plane of the ecliptic. In other words, in a chart, the ecliptic is the chart wheel itself. When we look at the positions of the planets along the ecliptic, we are looking at where along our journey we encounter these energies. The position of the Sun at birth shows the point where we chose to begin our quest for self-expression and self-realization in this lifetime. The ecliptic, the chart wheel, represents the actual course that we will follow.

Viewed in this way, the nodes represent the places in the chart where our past, our soul memories (the Moon), intersect with our current conscious experiences and our current cycle of lessons and growth. The South Node is the point where we are able to dip below the ecliptic into our past and access our memories from previous journeys. The North Node is the point where our past lessons intersect with our present journey, the point where we emerge from the past and move into new territory. It is the point where our past lessons come up into the light of conscious awareness and enable us to look at this piece of our past in an entirely new way.

Since the South Node relates to our past soul lessons, and the North Node relates to our current soul lessons, the traditional association of the South Node to our past and our karma, and the North Node to our future, or dharma, makes sense.

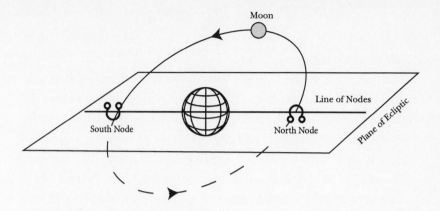

The Nodes Are Always in Perfect Opposition

Next, let's look at the opposition aspect. Planets are said to be in opposition when they occupy points across the wheel from each other, at an angle of 180 degrees. Traditionally, oppositions were considered "hard" or "challenging" aspects, but fortunately this opinion has been largely updated.

When I look at oppositions, the keywords I start with are *balance* and *perspective*. (Others also use *compromise*, but I find that limiting. To me, *compromise* means that each person has to give up something that they want in order to get something that they want. *Balance* just means that there is agreement and harmony.) The thing about oppositions is that both planets really "want" the same thing; they just approach it from different ends of the spectrum. If we can get each planet to see things from the other's point of view, then we can find that middle ground where they can work together and both get what they want. This process is made easier by the fact that, being directly across from each other, the two planets can "see" each other easily and are able to gain greater perspective on the big picture.

Putting It Together

So, the key to the nodal axis is to get the South Node and the North Node to work with each other, right? Not quite. The nodes are not planets; they are mathematical points in the chart. This doesn't make them any less important than a physical body; it just makes them a bit different to work with. The planets, remember, represent physical urges (for want of a better word) and drives that we all have. We can either choose to work with our Mars, for example, or to ignore it. Either way, we're going to be very aware of its presence in our chart. When Mars is activated by transit, or when transiting Mars triggers activity in our chart, we feel it. If we choose to become aware of it and own it, we can learn to use the energy in the most constructive way possible. If we don't, we're still going to experience it.

The nodes, on the other hand, have to do with the spiritual or soul lessons that we can encounter in this lifetime. The nodes are where the path of our soul development intersects with the path of our physical experiences. Since we're presently incarnated on the physical plane, that is where our focus naturally lies. If we don't choose to work with the nodes and their lessons on a conscious level, we are probably not going to be very aware of how they manifest.

In order to really work with and experience the nodes, we have to become consciously aware of our spirituality, of our soul connection and our connection to the universe. We have to accept that we came here with a lesson that we chose to work on in this lifetime, and we have to be ready to ask what that lesson might be. Are we learning it now? Of course. But the soul path lesson is far more subtle than a Mars transit.

The nodal axis is not our spiritual path in this lifetime. What it is, or rather, what it can be, is a spiritual compass, pointing us in the right direction. The nodes show us where our spiritual path and our soul lessons intersect with our physical path, and they are the points where we can most easily align with our spiritual path.

So with the nodes, it's not just about perspective and balance, it's about learning how to consciously work with and integrate the lessons, gifts, and experiences indicated by the North and South Nodes. The nodes can tell us what we have to work with, and show us what direction to go in to experience our lessons and to look for our true path in this lifetime.

The South Node

The South Node sheds light on our past. By its location in the chart, the South Node represents the types of experiences and memories that our soul wants us to have in this lifetime to help us on our developmental path. If our soul has gone to all of the trouble of bringing up these experiences, shouldn't we pay attention? And would a soul filled with love and light intentionally send us back into the material realm with only excess baggage? Of course not.

The South Node represents gifts to us from past lives. The lessons, skills, talents, and abilities that we struggled so many years to develop and master are still available to us through the South Node. The South Node is our soul's report card. It tells us what subjects we have passed: the South Node says, "Congratulations, you have passed Gemini Level I." For "incompletes" and less impressive marks, look to Saturn for further instruction and retesting, not to the South Node.

Just because we can now look at the South Node in a more positive light doesn't mean that the traditional warnings about the South Node being a "trap" don't have merit. Every planet and point in the chart has a highest potential and a potential trap. With the South Node, the trap is to mistake summer vacation for graduation. We may be able to take a break, but we're still in school. So long as we are incarnated on the Earth, we are here to learn.

Even though the South Node represents lessons and skills that we have learned, used, and often mastered in the past, it also is an indication that we still have more to learn about them in this lifetime. We may have learned the skills *too* well and limited our growth in other areas. We may have misused the information in the past, learning the letter of the lesson but not the spirit. Or it may merely be time to learn how to use the skills in a different way, to expand our mastery. In any event, for higher education, we look to the North Node.

The North Node

The South Node is not alone in being long overdue for a revised interpretation. True, the trap of the South Node is a tendency to stick with what is familiar, which can mean that we miss opportunities for growth and repeat old patterns. Also true, the way to avoid the trap of the South Node is to work with the North Node, which is why the North Node got such a great reputation. But the North Node also has a trap: the tendency to want to turn our back on the past in the single-minded pursuit of growth and new experiences.

The temptation of the North Node is to forget where we've been, and to focus only on where we are headed. The North Node, after all, brings success, happiness, abundance, luck, and freedom from the patterns and habits of the past. The North Node is the spiritual equivalent of the trip to the Bahamas that we've always wanted to take. And just like that trip to the Bahamas, the trap of the North Node makes us feel that much as we want it, we probably can't afford it. The North Node can seem to say, "You can get to the Bahamas, but you have to leave your nice, comfortable boat and swim there on your own. Oh, and by the way, there are probably sharks in the water." The trap of the North Node is just as dangerous as that of the South Node. We buy into the idea that we have to sever our ties to the past and create a new future. We dive into the water to swim out to the Bahamas, get part of the way there, and either get too tired or too scared of the sharks in the waters, and hurry back to the comfort and familiarity of our boat, the *S. S. South Node*. And if we've *really* bought into the trap, we may feel like a spiritual failure on top of it all.

Obviously, this is not an interpretation that is very supportive. My intention is not to bash the North Node, only to point out that the traditional interpretation of the nodes encourages us to stay away from the South Node to reap the rewards of the North Node, and that is just *not* what it is all about. The true process of the North Node is not about turning our back on the past. The North Node is about taking stock of the past, honoring it, working with it, building on it, and learning how to use it in a new way.

For example, the North Node Sagittarius/South Node Gemini axis doesn't tell us that we're done with Gemini and now have to learn Sagittarius. Instead, it tells us that we are now going to learn how to use our Gemini experiences in a Sagittarian way. It tells us that part of our path, part of our lesson in this lifetime, is to recognize that there *is* a point of balance between Gemini and Sagittarius, and finding it will be a key to our spiritual growth and development. By working with both the North Node *and* the South Node, we get to take the boat to the Bahamas instead of having to swim there.

Interpreting the Nodes

Before we can begin to integrate and interpret the nodal axis, there is one more factor to consider: the houses. The single most important thing to remember about the houses is that *the houses are NOT the same things as the signs.* Every book on the Moon's nodes that I have come across makes the same assumption and claim that, for example, the North Node in Aries is the same thing as the North Node in the 1st house. This is simply not the case. Remember that the signs are the roles the actors play and the costumes they wear, while the houses are the scenery, the locations, and the places where the actors go to play out their roles. The signs represent the underlying motivations and the evolutionary lessons that we must learn. The houses represent the areas of life where we are most likely to encounter these lessons. The signs, then, are the "What" and the houses are the "Where." The house placement of the nodes does not in any way change the fundamental lessons and gifts of the nodal axis; it simply shows where in our life we need to look to find the lessons and gifts.

The houses, too, have their lessons, and like the lessons of the sign axis, the house axis teaches the importance of balance and perspective. The opposing houses, like the opposing signs, represent areas of life that we must learn to integrate and harmonize. Just as we may tend to get a little too comfortable with the gifts of the South Node, we may also tend to focus more on the area of our life represented by the house of the South Node. We may also occasionally focus too much on the North Node or spend too much time involved in the affairs of the North Node's house, and forget to draw on the support and resources of our South Node. Because the houses represent the places where we will naturally encounter the nodes, if we want to work with our nodes, we can simply devote time to activities that relate to the houses in question, and then we will naturally encounter our nodes.

When interpreting the nodal axis, the first and most important factor to consider is the sign axis of the nodes. What do the two signs have in common? All opposing signs share some common theme; they simply approach it from different perspectives. Next, we look at the sign placement of the South Node and the North Node within this axis. What are the gifts that the South Node offers? What are the best and most wonderful expressions of the sign of the South Node? How can working with the North Node balance and enhance these gifts? By exploring these questions, we can discover some of the lessons of the specific nodal axis. Now, we can take this understanding and look at the house placement of the nodes to learn more about where in our life we will be able to experience and encounter these lessons.

Finally, let's try to put things into perspective. Remember that the nodes are not physical bodies, and because they are not physical bodies, they operate on a far more subtle level than the planets do. Unlike the planets, the nodes do not play a very big part in the development of an individual's personality. The nodes, however, are the key to understanding more about what our spiritual purpose is in this lifetime. Sometimes the nodes are closely connected with the planets in a chart, and the connection between the individual's physical life and his or her spiritual path is obvious. Sometimes the nodes seem to exist on their own, separate from the rest of the action in the chart. Whichever way the nodes appear to be linked with the other elements in the chart, the nodes can help us step back and see beyond the limitations of our time on Earth, and once again glimpse the bigger picture of the evolutionary journey of our soul.

2

THE HOUSE AXES

Before we look at the sign axes, let's take a moment to examine the houses in more detail. The most important thing to understand about the houses is that they are not the signs and they are not the planets. Houses represent different areas of experience, different sides of our personality that we encounter in different situations. When a planet is placed in a house, we have a predisposition to encounter the energy of that planet when involved in the affairs of that house. The sign on the cusp of a house, and signs intercepted in a house, color the tone of the affairs of that house.

For example, the 1st house is not the same as Aries. True, both are concerned with identity, but the 1st house is the *place* where we encounter the things that define our identity, and Aries is a sign that motivates planets to define an identity. Is there a correlation between the progression of the natural zodiac and the houses? Absolutely. They each represent an evolutionary cycle. The cycle of the houses, however, is more personal, and more experience-oriented, while the cycle of the signs is universal.

The next few chapters will interpret the different nodal axis combinations by sign and house. Although the houses are integrated in the interpretations, it's important to understand the dynamics of the various house axes on their own terms.

The Three Types of Houses:
Angular, Succedent, and Cadent

Houses can be either angular, succedent, or cadent. The angular houses (1, 4, 7, and 10) are the houses that are, not surprisingly, usually connected to the angles.[1] Because the angular houses are closest to the main "doors" to the outside world (the Ascendant, Midheaven, Descendant, and *Imum Coeli*), planets in angular houses tend to express in more external and obvious ways. Angular houses are very action-oriented and are similar in quality to cardinal signs. Succedent houses (2, 5, 8, and 11) follow (or succeed) the angular houses, and are relatively neutral in quality. Planets in succedent houses aren't in the foreground of the chart, but they're not in the background, either. Succedent houses are concerned with security and stability, and are similar to the fixed signs in this respect. The cadent houses (3, 6, 9, and 12) follow the succedent houses, and are the least expressive points in the chart. Planets in cadent houses have much greater difficulty making themselves noticed. Cadent houses are concerned with learning and with adaptation, and are similar to the mutable signs. The last few degrees of the cadent houses, however, usually the 4 to 6 degrees before the angles, are extremely strong. Planets in a cadent house within 6 degrees of an angle are considered to be angular.

The 1st House/7th House Axis

The 1st house/7th house axis is concerned with personal identity in relationship to others. In the simplest terms, it is the "I versus You" axis. The cusp of the 1st house, the Ascendant, is where we first entered into the physical world, and the 1st house is where we seek to form an identity and to begin to express it. In the 1st house we have no concept of our apparent separation from the source; we merely seek to discover who we are. Planets in the 1st house are fundamental to our formation of an identity, and the sign or signs in the 1st house, especially the sign on the cusp of the house, show our motivation for

---

1. Not all house systems associate the angles with the cusps of the angular houses. For more information on the different house systems, see chapter 6 of my book *Astrology: Understanding the Birth Chart*.

and method of forming and expressing our identity. The 1st house and the Ascendant form the mask that we wear when we go out into the world and relate to other individuals.

While the 1st house is the house of the self, the 7th house is the house of the "not self." In the 7th house we experience the first realization that there are other individuals in the world, each with as complex a set of needs and desires as our own. For the first time, we can define our own identity by exclusion and reflection. In the 7th house we seek social and intellectual action, or, perhaps more accurately, interaction, on a one-to-one basis. In the 7th house we use our relationships with other individuals to experience and encounter the qualities that we feel we lack in ourselves. The 1st house contains a yearning to be an individual and have an identity and, at the same time, an indefinable feeling of incompleteness. Something is missing in ourselves, and where we look to discover these missing pieces is the 7th house.

The challenge of the 7th house is to realize that everything we find there is really a part of ourselves. The 7th house is the mirror of the 1st house, and every quality that we seek and admire in a partner, every annoying habit that we hate in an enemy; all of these come from ourselves. What we are not ready to recognize or accept in the 1st house, we reflect back to ourselves in the 7th house, through others.

The lesson of this axis is perhaps one of perspective. What is in our 1st house is too close to us, and therefore we are too subjective about it. What is in our 7th house appears to be too distant from us, even to the point of being given to another person, and we can be too objective about it. As much as we must learn to "own" the planets in our 7th house, we must also learn to loosen our grasp occasionally on the planets in our 1st house.

The 2nd House/8th House Axis

The 2nd house/8th house axis is concerned with security. The 2nd house is concerned with material security and the 8th house is concerned with emotional and soul security.

In the 2nd house we find values, beliefs, structures, and physical objects that support, enhance, and reinforce our sense of identity that we created in the 1st house. The 2nd house contains everything that we can call "mine" in any sense of the word. These are the things that protect our identity, that sustain it. The 2nd house contains everything that we need in order to feel comfortable and secure with a physical identity and existence. It is where we impulsively feel the need to draw things to us to give us a sense of material safety in a physical incarnation.

The 8th house, by contrast, contains what we need to have our emotional and soul needs met by another individual. In the 7th house we discovered that the collective consists of other individuals, and we discovered the need to relate to, interact with, and make adjustments for another individual. In the 8th house we seek to have our emotional and soul needs met by another. The 8th house is where we first encounter the feeling that we are a part of something far greater than ourselves, and where we seek to make an emotional and soul connection to the cosmos through another.

Being directly opposite the 2nd house of what is "mine," the 8th house contains everything that is "not mine" and is generally associated with other people's money and resources. The 8th house is about more than money, though; it is also the house of death, taxes, and inheritance. Being a succedent house, the 8th house is concerned with security—in this case, with emotional and soul security. The 8th house is visible, being above the horizon in the southern hemisphere, and it involves other people because it's in the western half of the chart. Nevertheless, the 8th house is a very private room; it's where we keep many of our secrets as well as our fears. The 8th house relates to the occult and to buried and hidden things, so psychology and psychotherapy can be found in this house. Thanks to Freud and modern psychology, however, the 8th house has gradually become associated with sex (probably because of the Freudian connection between sex and death). Ancient astrologers would never have understood this— they knew that sex was supposed to be *fun* and therefore it belonged in the 5th house.

Just as the 2nd house contains the things that strengthen our sense of identity by being able to say, "These things are mine," the 8th house is where we may also strengthen our sense of boundaries between ourselves and others by saying, "These things are not mine; they are yours." The challenge with the 2nd and 8th houses is to find a balance between looking for the material security to keep us grounded in the physical plane and to help reinforce our illusion of separation from the universe, and the desire to dissolve the physical boundaries of the ego and identity and to discover the sense of belonging and emotional support that comes from merging with another person.

The 3rd House/9th House Axis

The 3rd house/9th house axis is concerned with exploring our relationship to our environment and learning how to make sense of it. In the 3rd house we first gain the realization that we are not what we grasped in the 2nd house: we are separate from our environment and from objects. Here we first encounter the "I versus It" distinction, and we begin to explore our immediate environment in order to make sense of it. In the 3rd house we develop the tools of language and communication. Once we discover that we are separate from our environment, we need ways to represent and communicate this distinction to ourselves.

The 3rd house, by extension, contains all the things in our immediate environment that we can contact without leaving our immediate sphere of experience. From this the traditional interpretation of brothers and sisters, short journeys, and communication begins to make more sense. In the 3rd house we develop our logical functions as well as our intellectual curiosity: there is so much to explore and to learn just in our immediate surroundings! Being a cadent house, the 3rd house is, of course, about learning. It also relates to religion and spirituality, although this seems to have more to do with religions that are not widely accepted at the time. (In classical astrology, the 3rd house was the house of heresy.)

The 9th house is where we begin to consciously adjust our concept of identity, questioning how we fit in, in the larger, social scheme of things. By the time we reach the 9th house, we have gotten used to the fact that there are other individuals in the world, and that we want to interact with them on a one-to-one basis. In the 9th house, however, we first realize that this will require that we adjust our sense of identity in a fundamental way. The 9th house is where we seek to explore the realms beyond our immediate environment. We seek to expand our awareness, our boundaries and our knowledge of the world in order to understand better who we are as individuals. The 9th house can contain many tools for this exploration, such as higher education, philosophy, religion, or travel to far-off lands. Opposite the 3rd house of early education and short journeys, the 9th house relates to higher education and long journeys—both in duration and in distance. The 9th house is where we explore the world, and experience and encounter new and unfamiliar cultures, ideas, and people. The 9th house is also related to organized religion and the clergy. We go to the 9th house when we dream and fantasize. What we study and learn in the 9th house is for our own personal growth and enlightenment.

The challenge of the 3rd house/9th house axis ultimately relates to the balance between the lower mind and the higher mind. The 3rd house, where we find our logical and analytical abilities, tends to focus on the smaller picture; by contrast, the 9th house, where we connect with our intuition and our imagination, looks for something bigger. By finding a balance between the 3rd house and the 9th house, we can both expand our understanding of the universe and apply this new knowledge on a daily basis.

The 4th House/10th House Axis

The 4th house/10th house axis is concerned with the issue of collective responsibility versus individual responsibility. The 4th house is where we take action and initiate on an emotional and a soul level, while the 10th house is where we take action on a material and physical level.

The cusp of the 4th house, the angle known as the *Imum Coeli*, or IC, is where our soul enters into our physical body. The 4th house is our connection to our past, our roots, and the emotional support and nurturance of the universe. Our family of origin is contained in the 4th house, as are our ancestors and things that nurture and protect us on an emotional level. The 4th house is where we seek to take responsibility for our soul connections, where we build structures to protect, nurture, and strengthen our connection to our family, our past, and our source. Although we can go here when we want to be alone, we generally go to the 4th house when we want to feel a sense of connection and community to those we love. And, of course, the 4th house is in the western hemisphere of the chart, so it is more focused on relationships and interacting with other individuals than it is on being alone and self-sufficient. The 4th house relates to real estate and property, and rules the father in the chart (even though modern astrology associates the mother with the 4th house). Even though the 4th house is an angular house, and therefore action-oriented, because it is the most hidden, least visible part of the chart, planets in the 4th house do not tend to express in very public ways.

The 10th house, on the other hand, is where we seek to establish material structures and tangible expressions of who we are as an individual. The 10th house, and particularly the Midheaven, the cusp of the 10th house, is the culmination of our physical manifestation and projection of our self into the material plane. The 10th house relates to our life path, which, if we're lucky, is also related to our chosen career. The 10th house is related to authority figures and policy makers and, traditionally, to that ultimate authority figure and policy maker, Mom. Even though modern astrology has decided that the father should be represented by the 10th house and the mother by the 4th house, in classical astrology, Dad is behind the scenes at the foundation, while Mom is the one who takes the more prominent role in shaping our life and how we appear in public. The 10th house relates to our career and life path in the sense that this is how we want to be recognized by society as an individual. While the 10th house is the most public and prominent part of the chart, it is also the most isolated and lonely. The roof is only big enough for one person at a time, and it's a long, narrow climb to get there. The angle associated with the 10th house, the Midheaven, is opposite the IC, where we are the least visible but the most connected to our roots and our source. The Midheaven is our crowning achievement as an individual, our most public face, and the way that we take responsibility for our role in society.

The challenge of the 4th house/10th house axis is to find the point of balance between our public life and our private life. We must learn how to draw strength and support from our roots, but we must also learn how to move beyond these roots and our past to define and express our unique identity as an individual.

The 5th House/11th House Axis

The 5th house/11th house axis focuses on our relationship as an individual to the group. In the 5th house we seek security for our identity in the group, while in the 11th house we seek social and intellectual security. This is the axis of group interaction.

In the 4th house we become aware of our emotional needs, and also that we may have to rely on the collective to support ourselves. In the 5th house we seek to earn this support and attention from the collective by being special. The traditional meanings of the 5th house all have this in common, and the 5th house contains everything that makes us feel special. Love affairs, romance, children, adventure, gambling, artistic creativity—all of these fill us with a sense of being different, of enjoyment and of pride in our individual and unique identity. In the 5th house we look for and find the ways that we support and become secure in our uniqueness.

In the 11th house we seek acceptance from a group of individuals as a rightful and equal member of the group. After the isolation of the achievement of the 10th house, we once again seek to integrate into a social support structure. We know that we are special; we have accomplished and expressed our unique individuality, and we do not wish to relinquish this. At the same time, we seek identification with others who are equally special. In this way, the 11th house contains groups of our own choosing, our peers, our friends, and social organizations to which we belong. In the 11th house we begin to shift our focus from our identity as an individual and to expand our focus to include our identity as a part of a collective. The 11th house is opposite the 5th-house game room, and the two do have much in common (they're both pretty fun places to be). The main difference is that when we're in the 11th house, we're spending time with groups of people; in the 5th house, we're either alone or in more intimate circumstances that usually don't call for more than one other person. The 11th house then relates to friendship and friends, and it also has to do with our hopes and wishes (as opposed to our dreams, which, you remember, are found in the 9th house).

The 5th house/11th house axis teaches us how to balance our need for personal expression with our need to be accepted and loved by others. In the 5th house we are only concerned with our own identity and with reminding ourselves of what it is that makes us so special and unique. In the 11th house we are less concerned about our individuality and more concerned about being a part of a group—the relationships with our friends become more important than our individual desires and interests. Balancing the two houses allows us to both give and receive love, to express our unique, creative identity, and to be appreciated for who we are.

The 6th House/12th House Axis

The 6th house/12th house axis is where we learn about becoming whole. The 6th house is a house of material learning and is where we learn what it takes to function as a part of the collective in a physical body on the physical plane. The 12th house is a house of emotional and soul learning and is where we learn what it takes to function as a part of the group soul on an emotional level.

When we reach the 6th house, we have exhausted our efforts to be accepted as a member of the collective because of how special we are. In the 6th house we attempt to gain acceptance through service and competence. The 6th house is where we really begin to explore the responsibilities of maintaining a physical body and a physical existence on a day-to-day basis. As our awareness of the collective continues to expand, as we move toward the realization that not only are we not alone, but that we are sharing this existence with other individuals, we begin to seek to improve our physical existence not merely for ourselves, but for others as well. The 6th house represents the material services that we perform to maintain and improve our physical and material membership in the collective. The 6th house is where we strive to earn our right to be manifest in a physical form.

The 12th house is where we learn and explore what it means to have a soul and an emotional body. The 6th house is concerned with the daily maintenance of our physical body; the 12th house is concerned with the daily maintenance of our soul and our emotions. In the 12th house we can discover true spirituality as opposed to the religion that we may encounter in the 9th house. In the 12th house we seek once again to join with the collective soul, to be free of the limitations of the physical.

The 12th house has been said to represent our unconscious, loss, and our ultimate undoing. In classical astrology, the 12th house is also associated with prisons and with being imprisoned. By extension, this is all accurate, although it has resulted in the 12th house being given a bad reputation. In the 12th house we seek to be free from our physical body and to once again merge with pure spirit. In the 12th house, therefore, our physical body is indeed a prison. In the 12th house we seek to lose our identity as an individual and to become one again with the universal life force.

The 6th house/12th house axis is about finding a balance between the needs of our physical body and the needs of our soul. We must take the time to nourish and recharge both, or our life will fall out of balance. Too much emphasis on the 6th house can result in a strong connection to the physical world, but a loss of our true spiritual connections. On the other hand, too much emphasis on the 12th house can result in strong spiritual ties, but difficulties in functioning in the material world.

3

THE ARIES/LIBRA
NODAL AXIS

The Aries/Libra nodal axis is the axis of identity. The purpose of this axis is to learn to develop a sense of self as an individual within the context of relating to other individuals. Both Aries and Libra seek to create a greater expression and develop a greater understanding of the self. While Aries defines the self through expression of individuality, Libra defines the self through relating with other individuals, and therefore by exploring and defining boundaries.

The function of Aries is to begin, to pioneer, and to create new life. Aries breaks away from the collective consciousness of Pisces when the infinite connection to everything becomes too limiting. In order to be able to continue to evolve, a part of the infinite separates and forms the illusion of an individual identity: this is the process of Aries. Aries is a trailblazer because it must push past all boundaries and shatter all limitations in order to express an individual identity. Aries is a leader because it is not comfortable being led or limited by others. Aries is courageous, inspirational, enthusiastic, original, and independent. However, Aries is entirely focused on expressing the self, and is therefore entirely ignorant of the presence of any other individuals and, more to the point, ignorant of the fact that its actions have repercussions and will affect and impact other people.

Libra, the opposite sign from Aries, is as naturally aware of other individuals as Aries is ignorant of them. Libra fully understands the responsibility of being an individual, and seeks to maintain balance and harmony in all aspects of one-to-one relationships. Ultimately, Libra is seeking to restore the balance between the self and the universe, the collective source of creation that we left when we formed our identity through the process of Aries. Through relating to and harmonizing with others, Libra is able to further define and strengthen our sense of personal boundaries. Libra is truly collaborative and can be charming, diplomatic, artistic, creative, objective, and entirely fair and impartial. Libra is very aware of personal responsibility and has a very strong sense of justice. This understanding of the burden of responsibility for our actions and the need for balance can become Libra's greatest challenge, however. When keeping the peace becomes the most important objective, Libra can sacrifice personal boundaries and individual needs in order to avoid a potential conflict. When every decision and every action will have a direct and equal reaction, not making a decision and not taking action can appear to be the safest way to maintain harmony. The result is that Libra can become either calculating and manipulative, or passive and reactive, denying its own individual needs.

The Aries/Libra nodal axis teaches that we must learn to develop our individual identity and yet maintain it in relationship to others. We must learn to develop appropriate personal boundaries, and maintain them in relationship to others. But at the same time, once those boundaries are defined, we have an obligation to ourselves to fully express our individual identity. We must learn to what degree we must be selfish and be allowed to express our individual needs and desires, and at what point we must learn to compromise and let the needs and desires of others take precedence in order to maintain balance and harmony in our relationships.

North Node Aries/South Node Libra

The South Node in Libra carries the gifts of balance, harmony, and beauty, and a fundamental appreciation for the finer points of relationships. Your skills at negotiating come from an almost natural ability to see both sides of any issue, in an entirely impartial light, and to be able to weigh the merits and consequences of each option. You also have a great appreciation for order, structure, and balance in all things. The trap of the South Node in Libra, however, is a tendency to want to maintain the harmony in relationships at all costs, and to deny all individual needs that could upset the dynamic of the relationship. The lesson here is to learn to work with the North Node in Aries, to explore and express your individual identity while still maintaining an awareness of the dynamics of interpersonal relationships.

The North Node in Aries teaches that it is not enough to simply have balance in a relationship: you must maintain balance while fully expressing your individuality at the same time. Learning to be impulsive and spontaneous is very important because this means that you are acting out of your own individual desires. If you do step on a few toes while you are discovering yourself, you can draw on your South Node gifts of charm and tact to smooth things over and restore harmony and a true balance to the relationship.

North Node in Aries in the 1st House/
South Node in Libra in the 7th House

With your South Node in Libra in the 7th house, you will encounter your South Node gifts of charm, diplomacy, and balance through your relationships. One of the challenges of the 7th house, however, is that we tend to give away planets in our 7th house, projecting them on others. You may feel that you lack the Libra qualities, and therefore you may tend to attract people into your life who embody them. Until you are able to accept the gifts of the South Node in Libra as being a part of you, you will experience and encounter them through your relationships. You may find that you have a pattern of giving away your power in relationships. This is the trap of the South Node in Libra in the 7th house. Your desire to maintain harmony and balance in your relationships may be so strong that you may tend to avoid making any decisions or taking any action that could possibly upset your partner. Often, this may involve not doing the things that you want to do, but rather always doing what others want.

Working with your North Node in Aries in the 1st house can help you balance this energy. However, we face a similar challenge with planets in the 1st house—while we tend to project the 7th house on others, we tend to overlook planets in the 1st house because they are such a fundamental part of who we are as individuals. The lesson of your North Node in Aries in the 1st house is to learn how to be selfish. You are already very aware of boundaries in relationships because of your South Node in Libra, and therefore you are also aware of your personal space. Being selfish means allowing yourself to actually fill up all of your space, and expressing yourself as an individual. By learning to make your own decisions and your own choices, and more importantly by expressing these choices, you will be asserting and defining yourself as an individual. You must continue to take advantage of your South Node gifts of diplomacy and balance, of course. The North Node in Aries in the 1st house is not about always insisting on getting your own way. However, as you learn to explore this energy, you will find it easier to participate actively in the give and take of relationships, and you will also learn how to maintain your own boundaries while in relationship.

North Node in Aries in the 2nd House/ South Node in Libra in the 8th House

With your South Node in Libra in the 8th house, you will encounter and experience your South Node gifts of charm, diplomacy, and balance when you are experiencing close emotional connections with other individuals. The 8th house is where we go to find a sense of emotional and soul security and self-worth. We seek a sense of inner peace through letting go of our physical and material attachments and merging with another individual on a deep, healing, and transformational level. The 8th house is also related to shared resources, and, like the 7th house, we often project our 8th-house planets on others, experiencing them through relationships. You will tend to be naturally aware of how to maintain balance in all of your 8th-house dealings. With respect to finances and shared resources, your impartiality and dedication to fairness make you extremely responsible. The trap of the South Node in Libra in the 8th house, however, is the tendency to become too dependent on others for your own emotional security. The need to maintain a relationship can lead to either self-sacrifice (where you forgo having your needs met in order to keep your partner happy) or to various forms of manipulation designed to keep your partner dependent on you.

The key to balancing this energy is to work with the North Node in Aries in the 2nd house. The 2nd house represents our personal resources; anything that we can call our own belongs in the 2nd house, and everything in the 2nd house helps reinforce our sense of individuality. Our possessions, resources, skills, and talents, as well as our physical body and senses, help define and support our identity as an individual. The North Node in Aries in the 2nd house challenges you to take a break from your relationships and your involvement with others, and to experience yourself as an individual, on your own terms. You must learn how to ground yourself, and how to give yourself permission to be selfish. Maintaining shared resources is fine, but you must also allow yourself to maintain your own boundaries and accept that some things belong only to you. The 2nd house is where we can ground ourselves in the physical realm, and one way of exploring this house is to spend time in nature, connecting to the Earth. By asserting your own self-worth, by expressing and exploring your personal resources, you will find the strength to maintain your individuality while you relate to others. More than that, you will discover where your boundaries need to be in order for you to share yourself with others, and not lose yourself in the process.

North Node in Aries in the 3rd House/
South Node in Libra in the 9th House

With your South Node in Libra in the 9th house, you encounter your gifts of charm, diplomacy, and balance on a more philosophical and intellectual level. The 9th house is where we expand our understanding of the world and how we relate to it as an individual, and it includes philosophy, higher education, religion, and long journeys. Your South Node gifts may indicate an affinity for understanding other people's views and perceptions of the universe, as well as the ability to discuss and debate about these often touchy subjects without causing offense. Libra's desire for balance and inherent ability to see both sides of any issue can indicate that you are truly open to new and different ideas. The trap of the Libra South Node in the 9th house, however, is that your desire to harmonize with the views and philosophies of others may prevent you from forming your own opinions and discovering your own point of view.

The key to balancing this energy is to work with the North Node in Aries in the 3rd house. The 3rd house relates to our familiar environment, and to how we communicate, reason, and make connections within that environment. The 3rd house also relates to our individual spiritual practices and beliefs—the ways that we connect to our spirituality outside of the structures of organized religion. The lesson and the challenge here is to choose a philosophy, to take one approach that you have discovered through your South Node in Libra in the 9th house, and apply it to your day-to-day life and see if it actually supports and enhances your sense of identity. You must learn to take the theoretical and universal, and make it practical and personal. Through these experiments, you will find that some approaches and ideas work for you, and some don't; but in every case, you will be able to form your own opinion based on your own personal experience. Just as important is learning how to express and communicate these experiences and opinions, particularly when you are relating to others and learning about their approach and philosophies of life. You will still be able to explore and debate alternative philosophies and new ideas with your South Node in Libra in your 9th house; however, by developing your North Node energy, you will be able to maintain your own identity and your own ideas rather than feeling obligated to adopt and adapt to those of others.

North Node in Aries in the 4th House/
South Node in Libra in the 10th House

With your South Node in Libra in the 10th house, you encounter your South Node gifts of charm, diplomacy, and balance as a part of your public self. The 10th house is where we seek to create a tangible manifestation of our individual identity. It is where we want to make our mark on the world, and where we make our contributions to society as an individual. The 10th house is often associated with our career, but more accurately it is our life path, which is often quite different from our job. Your talents and skills with relationships may often assist you with your career, and you may also feel the need to create beauty and harmony on a much larger scale. You may find that your job, your responsibilities, or your social and societal obligations make great demands on you, and this can lead you to the trap of the South Node in Libra in the 10th house. You may find that you devote so much time and energy to maintaining balance and communication with your public life that you have little time left for your own personal needs, or for the needs of your family and friends. Your sense of identity may eventually be defined primarily by how others see you; your public persona may threaten to consume your true identity, although you will probably not be able to see things this clearly. Your perception will probably center around the belief that you have certain responsibilities to your public; that other people are counting on you and that your perceived obligations to them outweigh any personal issues you may have.

The key to balancing this energy is to work with the North Node in Aries in the 4th house. While the 10th house is the most public part of the chart, and also the most focused on individuality, the 4th house is the most private part of the chart, and the area where we are the most connected to others. The 4th house is our home; it is where we connect with our families (both of origin and of choice); but more than this, the 4th house represents our roots, our past, our ancestry, and our tribal heritage. It is the foundation on which we build our individuality and our public self. With your North Node in Aries in your 4th house, you must learn how to go off and do things on your own, to take responsibility only for your own personal needs and to momentarily forget about any obligations you may have to others. You must take time to do things and be with people where you can express yourself without concern for how others will perceive you. Working with your North Node in Aries in your 4th house will help you remember that your public persona is only a mask; so long as you remember what your true face looks like behind the mask, then the mask can be a valuable tool for you.

North Node in Aries in the 5th House/
South Node in Libra in the 11th House

With your South Node in Libra in the 11th house, you encounter your South Node gifts of charm, diplomacy, and balance when you socialize with your friends and peers. The 11th house is where we seek a sense of social and intellectual security. In the 11th house we want to belong to a social community. Within this community, we learn appropriate behavior standards, we participate in group creativity, and we learn how to receive and accept the love of others. The focus of your need for relationship is apt to be less centered on other individuals and more on your relationship with your social circle as a whole. You are likely to be able to participate in group activities and to be able to contribute to the overall creativity of the group with ease, and your charm and tact go a long way towards diffusing any potential imbalances within the group. On a fundamental level, you are aware of your responsibility to the group as an individual; you understand that everything that you do has an impact on the group as a whole, so you are able to take the needs of the group into consideration before you act within the group. This, however, can lead you to the trap of the

South Node in Libra in the 11th house, which involves giving up your own personal identity in order to maintain a balance between yourself and the group. You must be careful that you do not become so dependent on the approval and acceptance of your friends that you sacrifice your own individuality. As always with the trap of the Libra South Node, you may tend to define yourself in terms of how others see you. In this case, you may find that your sense of self-worth comes from how well you are accepted and appreciated by your peers.

The key to balancing this energy is to work with the North Node in Aries in the 5th house. The 5th house is where we search for security in our identity as an individual. Everything that we can do that makes us feel special and unique, that makes us feel like we deserve the attention and acknowledgment of others, is found in the 5th house—that is why the 5th house is associated with such a mixed bag of concepts including children, the arts, gambling, and love affairs. With your North Node in Aries in your 5th house, you must learn to express your personal creativity in a spontaneous and impulsive manner. Create something that comes from you, something that you can be proud of, and don't worry if your friends will appreciate it. Part of the lesson of the North Node in Aries in the 5th house is to learn how to do things that you enjoy doing, and more than that, to do them on your own. Your South Node in Libra in your 11th house will ensure that you don't stray too far from the bounds of acceptable behavior; but you must also be able to act without even wondering what others will think or if others will appreciate what you do. Of course, this may feel threatening because your South Node in Libra in the 11th house wants to be loved by others. The final part of the lesson of the North Node in Aries in the 5th house is that you must learn to give love freely, impulsively, and honestly. When you share yourself with others, when you learn to give love without caring if it is returned, you will receive love as well.

North Node in Aries in the 6th House/
South Node in Libra in the 12th House

With your South Node in Libra in the 12th house, you may not be consciously aware of your gifts of charm, diplomacy, and balance, even though they are very evident to other people. The 12th house is like our shadow: we can't see it directly because we're usually facing the light (of consciousness, that is), but it's very visible to everyone else. Our awareness of our 12th house seems to come from our unconscious and our subconscious. You have an unconscious and intuitive understanding of relationships and of your responsibility as an individual. The 12th house brings out the spiritual side of Libra; Libra is about one-to-one relationships, of course, but this can also mean exploring the one-to-one relationship between ourselves and our spiritual beliefs. You have the opportunity to discover balance and harmony on a spiritual level, and to experience a fulfilling and peaceful spiritual life. The trap of the South Node in Libra in the 12th house is that you may tend to sacrifice your individuality in the name of spirituality. Spending too much time in the 12th house can mean that we lose our connections to the physical reality; we retreat so far into ourselves that we lose touch with everything outside of ourselves. And the fact is that even though we *are* multidimensional, eternal souls, we are currently having a human experience, and that means that we have to learn to experience our spiritual connections and awareness in the context of living on the material plane.

The key to balancing this energy is to work with the North Node in Aries in the 6th house. While the 12th house relates to our emotional and soul needs, the 6th house relates to our routine physical needs. The 6th house is related to illness because if we're not taking care of our body with proper nutrition, rest, and exercise on a daily basis, we will get sick. The 6th house is also related to our job and work environment, which is an essential part of maintaining our physical existence. The North Node in Aries in your 6th house reminds you to come out of your isolation and to come back to the physical world; more than that, it tells you to take action on the physical plane. Your South Node in Libra in your 12th house may encourage you to give selflessly and to serve the greater spiritual good. The North Node in Aries in the 6th house teaches that you must learn to serve through your actions on the material plane. You must learn how to carry your spiritual connections with you into your daily life. The 6th house and the 12th house share the common ideals of service; the difference is that the 12th house tends to incline towards self-sacrifice while the 6th house is more about being of use to others in a practical sense. The challenge for you is to take the spiritual ideals that you have with your South Node and translate them into practical guidelines that you can express and embody in the physical through your daily routines. The more you are able to express your individuality through your service to your spiritual ideals, the stronger your connections will become to both the spiritual and the physical.

North Node in Aries in the 7th House/
South Node in Libra in the 1st House

With your South Node in Libra in the 1st house, your gifts of charm, diplomacy, and balance are very much a part of your sense of self; you access them naturally and may not even be consciously aware of them. The 1st house contains everything that we identify as being a fundamental part of who we are. In fact, it's quite difficult to gain perspective on the contents of the 1st house because it's too close to us. You may tend to define yourself in terms of how you relate to other individuals. The trap of the South Node in Libra in the 1st house is that since all of Libra's energy is focused on relationship, and that Libra defines the self in terms of others, you may have difficulty discovering who you are as an individual and expressing this. You may find that you rely on your relationships and interactions with others to define your personal boundaries. If you are not actively relating to others, you may not be able to define yourself as an individual.

The key to balancing this energy is to work with the North Node in Aries in the 7th house. The 7th house is the house of one-to-one relationships, and the biggest challenge of the 7th house is that we tend to give away the planets in this house, projecting them on others and experiencing the energies and lessons of these planets through our relationships. Working with the North Node in Aries may be a challenge, because you may feel that you lack the qualities of Aries. As a result, you will tend to attract people to you who embody these qualities. These relationships will help you learn how to accept and integrate the lessons of the North Node in Aries in the 7th house. You may find that you tend to attract and be attracted to assertive, independent and self-motivated individuals; individuals who, in other words, embody the Aries energy of your North Node. Rather than falling into the trap of your South Node in Libra and allowing these individuals to make all of the decisions in the relationship, your lesson is to discover your own sense of individuality through these relationships. The gift of your South Node in Libra in your 1st house is an awareness of boundaries in relationships. The lesson of your North Node in Aries in your 7th house is to learn how to expand to fill up your allotted space, to be able to express yourself freely and impulsively, and to maintain balance and harmony in your relationships with other individuals who are also freely and fully expressing themselves.

North Node in Aries in the 8th House/
South Node in Libra in the 2nd House

With your South Node in Libra in the 2nd house, your gifts of charm, diplomacy, and balance are important skills and resources to you. The 2nd house represents our personal resources; anything that we can call our own belongs in the 2nd house, and everything in the 2nd house helps reinforce our sense of individuality. Our possessions, resources, skills, and talents, as well as our physical body and senses, help define and support our identity as an individual. Your ability to form partnerships, to negotiate, to reason, and to communicate are the foundation of your ability to generate material wealth and success. The Libra influence in your 2nd house also indicates that you have a great appreciation for beauty in all things, and may also enjoy all forms of art, both as a patron and as a creator. Relationships can be a great resource to you; however, they can also become too important to you. The trap of the South Node in Libra in the 2nd house is that your sense of self-worth may be based on what you believe others think of you. You may tend to view yourself and your resources in terms of your relationships, and come to depend on others to provide you with feedback on your value as an individual. This may manifest as a need for constant approval and validation from other people.

The key to balancing this energy is to work with the North Node in Aries in the 8th house. The 8th house is where we go to find a sense of emotional and soul security and self-worth. We seek a sense of inner peace through letting go of our physical and material attachments and merging with another individual on a deep, healing, and transformational level. The 8th house is also related to shared resources, and, like the 7th house, we often project our 8th-house planets on others, experiencing them through relationships. The North Node in Aries in the 8th house encourages you to explore your own individual needs and desires within the context of your relationships. In the 8th house, we find emotional validation and support through connecting with another individual. The lesson here is to learn how to assert your own individuality while maintaining your relationships to others. Aries energy is impulsive and pioneering; you must learn how to move through your fear of rejection and offer yourself completely to another, trusting that he or she will do the same in return, and you will be able to experience the healing power of this type of union and rebirth. The 8th house is also related to things that are hidden or buried beneath the surface. One of the lessons of the North Node in Aries in the 8th house is to learn how to explore the unknown, to look beyond appearances and find the core emotional truth. Although this certainly applies to relationships, it can also relate to exploring your own unconscious needs.

North Node in Aries in the 9th House/
South Node in Libra in the 3rd House

With your South Node in Libra in the 3rd house, your gifts of charm, diplomacy, and balance will tend to manifest in your ability to communicate and to experience your immediate environment. The 3rd house represents the world that is most familiar to us, and the ways that we make sense of the world through logic, reason, and language. The 3rd house also relates to our individual spiritual practices and beliefs—the ways that we connect to our spirituality outside of the structures of organized religion. You may find that you have a gift for clear and logical communication, and that you are able to maintain balance and harmony in your world through applying your rational mind. Libra's gifts of diplomacy are quite evident in the 3rd house, and you may find that you tend to be the peacemaker of your neighborhood. The South Node in Libra in the 3rd house not only has the ability to see both sides of any issue, but more importantly has the ability to understand and communicate the different points of view to everyone involved. The trap of the South Node in Libra in the 3rd house, however, is the tendency to adhere to the limitations of the rational mind. Many problems can be solved through logic and reason; but many problems require breaking the rules, moving outside of the rigid boundaries of the familiar and the known, and gaining a more cosmic perspective.

The key to balancing this energy is to work with the North Node in Aries in the 9th house. The 9th house is where we expand our understanding of the world and how we relate to it as an individual, and it includes philosophy, higher education, religion, and long journeys. Part of the challenge of your North Node in Aries in your 9th house is to allow yourself to put your own personal stamp on things; to discover your own individual philosophy, and to be inspired by your own personal understanding of the universe and how you relate to it. Incorporating this new understanding into your daily life will be quite easy, thanks to your South Node. And what you seek to communicate and share with others through your South Node in Libra will be that much more valuable when it originates from your higher inspiration rather than from your rational mind.

North Node in Aries in the 10th House/
South Node in Libra in the 4th House

With your South Node in Libra in the 4th house, your gifts of charm, diplomacy, and balance are closely related to your private life and family connections. The 4th house represents our home, our families (of choice and of origin), and our connection to our past through our ancestors and tribal heritage. The 4th house is our foundation—the rock on which we build our individual life. It supports us, and through the foundation of the 4th house, we are able to achieve the public, individual accomplishments of the 10th house. Your sense of self is naturally very connected to your roots, and maintaining a balanced relationship with your family, and a personal relationship to your roots and your heritage, is very important to you. You have a natural awareness of your personal responsibility to your family. The trap of the South Node in Libra in the 4th house, however, is that your family ties may begin to limit and restrict you rather than support and encourage you. Rather than being free to express your own individuality, you may feel that you need the approval of your family; you may find that you are trying to be the person you think your family wants you to be rather than being the person you truly are as an individual. You may fear that if you expressed yourself freely, that you would upset the balance and harmony of your family, and risk losing the emotional connections and support. The irony is, of course, that if you are not able to express yourself as an individual, you are also not able to truly feel the support and emotional connections with your family, because you are not able to be open and honest with them.

The key to balancing this energy is to work with the North Node in Aries in the 10th house. The 10th house is the most public part of the chart. It relates to our career and our life path; it is where we as individuals find our most significant accomplishments. You can learn to experience your own individuality through actively pursuing your public life. Taking autonomous action, making your own decisions, or even being your own boss can help you reach this goal. You can certainly draw on the support of your family; your private life is, after all, your foundation. But you must feel free to pursue your own personal dreams, to take your own personal risks and enjoy your own successes and failures on your own terms. Part of the challenge of the North Node in Aries in the 10th house is learning when you must put your own desires and your responsibilities to the outside world ahead of your responsibilities to your family.

North Node in Aries in the 11th House/
South Node in Libra in the 5th House

With your South Node in Libra in the 5th house, you encounter your gifts of charm, diplomacy, and balance whenever you explore and express your personal creativity. The 5th house is where we search for security in our identity as an individual. Everything that we can do that makes us feel special and unique, that makes us feel like we deserve the attention and acknowledgment of others, is found in the 5th house—that is why the 5th house is associated with such a mixed bag of concepts including children, the arts, gambling, and love affairs. Because Libra is often associated with the fine arts, including painting and music, you may find that you enjoy expressing yourself and exploring your creativity in this manner, and sharing your creations with others. However you express your creativity, you are likely to do so with an eye towards what you think will be the most aesthetically pleasing to others. The trap of the South Node in Libra in the 5th house is that you may become so focused on impressing and pleasing others with your creativity that you may begin to lose your sense of individuality. The 5th house also relates to the love that we give to others. Although you are able to express love fairly and unconditionally, the trap of the South Node in Libra in the 5th house may encourage you to always give in your relationships out of the fear that if you ever stop giving of yourself, if you ever stop trying to please others, then you will not receive any love in return.

The key to balancing this energy is to work with the North Node in Aries in the 11th house. The 11th house is where we seek a sense of social and intellectual security. In the 11th house we want to belong to a social community. Within this community, we learn appropriate behavior standards, we participate in group creativity, and we learn how to receive and accept the love of others. With your North Node in Aries in your 11th house, you must learn to assert your individual identity in social situations. Working with this energy requires impulsive action and self-expression, which may scare your South Node in Libra in the 5th house, because if you break out of your natural pattern of wanting to please others, you run the risk of not receiving the attention and approval for your creativity that you crave. By asserting your individuality, however, you will naturally gravitate towards groups of friends and peers that truly support and accept you for who you are, and who will be able to appreciate and honor you for your creativity and uniqueness. Perhaps the hardest lesson of the North Node in Aries in the 11th house is learning to allow yourself to receive the love and appreciation of others. Once you discover the balance between maintaining a relationship with your friends and peers and asserting and expressing your individuality within those relationships, you will also discover the balance between the love that you give to others and the love that you receive from them in turn.

North Node in Aries in the 12th House/
South Node in Libra in the 6th House

With your South Node in Libra in the 6th house, you encounter your gifts of charm, diplomacy, and balance in your daily routines. The 6th house contains everything that we must do on a daily basis to maintain our physical existence. The 6th house is related to illness because if we're not taking care of our body with proper nutrition, rest, and exercise on a daily basis, we will get sick. The 6th house is also related to our job and work environment, which is an essential part of maintaining our physical existence. You may find it very easy to maintain a sense of harmony in your everyday routine. The relationships you have with the individuals you meet on a daily basis, particularly with your co-workers, may be quite important to you, and you are likely to be very skilled at keeping the lines of communication open and maintaining a pleasant and enjoyable work environment. In addition to our daily routines and responsibilities, the 6th house is where we look to serve others, to assist in making the world a better place. Because Libra operates primarily in the realm of one-to-one relationships, you may find that you tend to want to help the individuals you work with, and to do your best to make their lives easier and more enjoyable. The trap of the South Node in Libra in the 6th house is that you may tend to give up your individuality and forego your own needs in order to be of service to others. You may find that you try to maintain relationships by making yourself useful. Taken to the extreme, this can mean that you tend to take a subservient role in relationships that should be between equals.

The key to balancing this energy is to work with the North Node in Aries in the 12th house. While the 6th house relates to how we take care of our physical health, the 12th house is how we take care of our spirit and our soul. In the 12th house we encounter our need to feel a part of a spiritual and emotional community; to let go of our individuality and to merge once again with the universe. We retreat to the 12th house whenever we need to take a break from the demands of our daily routines. Exploring the 12th house can also mean exploring your spirituality and searching for your spiritual identity. By working with your North Node in Aries in your 12th house, you can begin to remember that we are truly multidimensional, eternal souls, and that although we are currently having a human experience and spending time on the physical plane, we are not limited by our physical experiences. Your South Node in Libra in your 6th house means that you will always be aware of your responsibilities and duties while on the physical plane. But letting yourself freely explore the spiritual can help you maintain your perspective, not to mention your individuality. Being of service to others on the physical plane is certainly important; but sometimes you have to learn how to be of service to yourself first.

When you learn to maintain an understanding of your true spiritual identity as an eternal, multidimensional being that is part of all of creation, you will be able to give of yourself without giving up yourself.

North Node Libra/South Node Aries

With the South Node in Aries, expressing your individuality is something that comes quite easily to you. The gifts of the Aries South Node include decisiveness, leadership, and a passion for life. Although the South Node in Aries is not directly related to leadership, the pioneering and trailblazing spirit that comes with the South Node in Aries often means that where you lead, others will follow. Making decisions comes very easily to you, and you are used to taking immediate action once you've made up your mind. The trap of the South Node in Aries, however, is being ignorant of the needs and boundaries of others. You may become so self-aware and so self-centered that you are either unaware or unconcerned with how your actions impact other individuals. You may either tend to be a loner, avoiding relationships because you don't want to have to compromise your own desires, or you may tend to automatically become a leader, assuming that others will either fall in step with you or else get out of the way.

Working with the North Node in Libra will teach you how to balance your individual identity with the other individuals in the world. Through relationships, you can learn that you end where others begin and become more consciously aware of your individual boundaries, which in turn reinforces your sense of individual identity. The North Node in Libra also teaches how to take responsibility for how your actions affect other individuals. When you become aware that every action you take has repercussions and that you are responsible, to one degree or another, for those repercussions, you may find that you tend to think a bit longer and more thoroughly before you act. As you relate to others, you will learn that you can maintain balance and harmony in your relationships without compromising your individuality.

North Node in Libra in the 1st House/
South Node in Aries in the 7th House

With your South Node in Aries in the 7th house, you will encounter your South Node gifts of autonomy, individuality, and spontaneity through your relationships. One of the challenges of the 7th house, however, is that we tend to give away planets in our 7th house, projecting them on others. You may feel that you lack the Aries qualities, and therefore you may tend to attract people into your life who embody them. Until you are able to accept the gifts of the South Node in Aries as being a part of you, you will experience and encounter them through your relationships. Rather than acknowledging your South Node in Aries gifts as your own, you may feel that you lack the ability to be autonomous and self-motivated, and instead attract people into your life who are extremely confident and sure of themselves. These individuals, however, may also embody the trap of the South Node in Aries in the 7th house and not respect your personal boundaries or take your individual desires or feelings into account when they act. They may appear to be selfish, arrogant, and completely oblivious to the fact that the world does not revolve around them. Of course, you'll only experience these individuals in this way if you yourself have fallen into the trap and are in fact behaving in this way towards others.

Working with your North Node in Libra in the 1st house can help you balance this energy. However, we face a similar challenge with planets in the 1st house—while we tend to project the 7th house on others, we tend to overlook planets in the 1st house because they are such a fundamental part of who we are as an individual. The lesson you must learn from your North Node is perhaps not the most obvious one. Because of your natural tendency to give away your South Node in Aries energy in relationships, you are already quite aware of how to maintain balance and harmony in your relationships. What you must learn from your North Node in Libra in your 1st house is that the only way to truly find balance is to express your individuality completely, while others do the same. Balance and harmony are not attained by automatically deferring to the needs and desires of others, but rather by expressing your own individuality and your own desires fully, and with the awareness that when you meet others as equals, you will then find fair and just ways to compromise when necessary.

North Node in Libra in the 2nd House/
South Node in Aries in the 8th House

With your South Node in Aries in the 8th house, you will encounter and experience your South Node gifts of autonomy, individuality, and spontaneity when you are experiencing close emotional connections with other individuals. The 8th house is where we go to find a sense of emotional and soul security and self-worth. We seek a sense of inner peace through letting go of our physical and material attachments and merging with another individual on a deep, healing, and transformational level. The 8th house is also related to shared resources, and, like the 7th house, we often project our 8th-house planets on others, experiencing them through relationships. The 8th house also relates to our individual journey through our shadow self, where we seek to confront, understand, and ultimately integrate our demons. This is often a shared process, and with your South Node in Aries in your 8th house, you have the ability to initiate emotional connections with others. You are also confident and self-assured enough to share your resources with a partner on a material, emotional, and spiritual level. The trap of the South Node in Aries in the 8th house, however, is that you may tend to be more self-centered in your relationships, being primarily concerned with your own journey and transformation, and losing sight of the fact that you're making that journey with a partner—a partner who also has emotional needs.

The key to balancing this energy is to work with the North Node in Libra in the 2nd house. The 2nd house represents our personal resources; anything that we can call our own belongs in the 2nd house, and everything in the 2nd house helps reinforce our sense of individuality. Our possessions, resources, skills, and talents, as well as our physical body and senses, help define and support our identity as an individual. The lesson here is to learn to view your relationships as a valuable resource, one that reinforces your sense of self as an individual. The North Node in Libra in your 2nd house also shows that you can ground and center yourself through the arts; being in the presence of great beauty, balance, and harmony can help you see the beauty and harmony in your own life.

North Node in Libra in the 3rd House/ South Node in Aries in the 9th House

With your South Node in Aries in the 9th house, you encounter your gifts of autonomy, individuality, and spontaneity on a more philosophical and intellectual level. The 9th house is where we expand our understanding of the world and how we relate to it as an individual, and it includes philosophy, higher education, religion, and long journeys. Your South Node gifts may tend to lend a decidedly individualistic approach to these pursuits, as they may be closely associated with your drive to define and express your individual identity. The independent spirit of Aries will tend to be drawn more to an individual spiritual quest than to the group dynamic of organized religion or philosophies. The trap of the South Node in Aries in the 9th house, however, is that the Aries need to pioneer and to not be controlled, limited, or influenced by others can make your spiritual journey a focused one that has no direction or guidance. The ideas and concepts that you explore and embrace may have little practical application in your life and therefore will be ultimately of little value to you.

The key to balancing this energy is to work with the North Node in Libra in the 3rd house. The 3rd house relates to our familiar environment, and to how we communicate, reason, and make connections within that environment. The 3rd house also relates to our individual spiritual practices and beliefs—the ways that we connect to our spirituality outside of the structures of organized religion. The North Node in Libra in the 3rd house teaches the importance of finding a balanced application of the spiritual lessons learned in the 9th house. It's not enough to define your own philosophy and spiritual path because you must also be able to live with and relate to other individuals. While you do have your own unique place in the universe, you're not alone in the universe. You must learn how to balance your beliefs and ideals with other individuals, and to integrate these beliefs and ideals in your daily life and relationships.

North Node in Libra in the 4th House/
South Node in Aries in the 10th House

With your South Node in Aries in the 10th house, you encounter your South Node gifts of autonomy, individuality, and spontaneity as a part of your public self. The 10th house is where we seek to create a tangible manifestation of our individual identity. It is where we want to make our mark on the world, and where we make our contributions to society as an individual. The 10th house is often associated with our career, but more accurately it is our life path, which is often quite different from our job. You may find that you are a natural leader in business situations. You are confident and able to make your own choices and decisions, and don't expect to take direction from anyone else. Because Aries energy doesn't like to wait around, it tends to be a trailblazer and is quite often associated with leadership ability. However, a big part of your spiritual lessons with your nodal axis relates to when it is appropriate for you to act on your own, and when it is more appropriate for you to take others into consideration. In business situations, when competition is ruthless, it's expected that you will step on some toes. This approach to relationships, however, is not appropriate in your personal life. The trap of the South Node in Aries in your 10th house is to assume that you can always do as you please in all areas of your life.

The key to balancing this energy is to work with the North Node in Libra in the 4th house. While the 10th house is the most public part of the chart, and also the most focused on individuality, the 4th house is the most private part of the chart, and the area where we are the most connected to others. The 4th house is our home; it is where we connect with our families (both of origin and of choice); but more than this, the 4th house represents our roots, our past, our ancestry, and our tribal heritage. It is the foundation on which we build our individuality and our public self. With your North Node in Libra in the 4th house, you must learn to experience and express harmony, balance, and diplomacy in your personal life and familial relationships. Your personal life cannot be run in the same way as your professional life. While this doesn't mean that you should compromise your individuality, it does mean that you must learn to take others into account, and to search for a point of balance, wherein everyone is happy and feels that they are making equal contributions to the whole.

North Node in Libra in the 5th House/
South Node in Aries in the 11th House

With your South Node in Aries in the 11th house, you encounter your South Node gifts of autonomy, individuality, and spontaneity when you socialize with your friends and peers. The 11th house is where we seek a sense of social and intellectual security. In the 11th house we want to belong to a social community. Within this community, we learn appropriate behavior standards, we participate in group creativity, and we learn how to receive and accept the love of others. You may find that you have a natural tendency to be the leader of your peer group. You may be the one who suggests the group activities, if only because if the group wants to do something else, the South Node in Aries is more likely to go off on its own than to stick with the group. The trap of the South Node in Aries in the 11th house, of course, is to become so determined to follow your own impulses that you either no longer have a peer group to share activities with, or you become so much the leader of the group that the possibility for discussion and a true peer group dynamic no longer exists. The South Node in Aries in the 11th house needs to be recognized as an individual by the others in the group, but not at the expense of the group itself.

The key to balancing this energy is to work with the North Node in Libra in the 5th house. The 5th house is where we search for security in our identity as an individual. Everything that we can do that makes us feel special and unique, that makes us feel like we deserve the attention and acknowledgment of others, is found in the 5th house—that is why the 5th house is associated with such a mixed bag of concepts including children, the arts, gambling, and love affairs. With your North Node in Libra in the 5th house, you must learn how to find balance and harmony in your creative expression. Even though the 5th house tends to be focused more on the individual, the Libra energy of relationships and balance is very prominent for you in your 5th-house activities. Ultimately, you can find that what makes you feel the most special as an individual is being accepted and acknowledged by another individual. And by focusing on finding the balance in one-to-one relationships, you can also find it easier to manage the group dynamic and avoid the traps of the South Node in Aries in the 11th house.

North Node in Libra in the 6th House/
South Node in Aries in the 12th House

With your South Node in Aries in the 12th house, you may not be consciously aware of your gifts of autonomy, individuality, and spon-taneity, even though they are very evident to other people. The 12th house is like our shadow: we can't see it directly because we're usually facing the light (of consciousness, that is), but it's very visible to every-one else. Our awareness of our 12th house seems to come from our unconscious and our subconscious. You may have an unconscious need to act out and express your individuality and your individual identity. Since the 12th house is where we encounter our need to feel a part of a spiritual and emotional community, and to let go of our indi-viduality and merge with the universe, this can be a somewhat uncon-ventional energy to experience. You are apt to take a private approach to your spiritual path, preferring to spend time alone getting to know yourself better. The trap of the South Node in Aries in the 12th house, however, is the tendency to become too isolated—to retreat entirely into the cloister of the 12th house and to put too much emphasis on exploring and expressing your individual spiritual identity.

The key to balancing this energy is to work with the North Node in Libra in the 6th house. While the 12th house relates to our emotional and soul needs, the 6th house relates to our routine physical needs. The 6th house is related to illness because if we're not taking care of our body with proper nutrition, rest, and exercise on a daily basis, we will get sick. The 6th house is also related to our job and work environment, which is an essential part of maintaining our physical existence. The North Node in Libra in the 6th house reminds you to come out of your isolation and interact with other individuals. Libra energy is all about relationships and about balance. On a more general level, your lessons are to learn how to balance your spiritual life with your daily life; but on a more specific level, your goal is to learn how to express the spiritual identity that you can discover through your South Node in Aries in the 12th house, and to maintain that energy while relating to other individuals in your day-to-day life. You must find the courage to work through the challenges in your relationships and avoid the temptation to retreat into isolation and to hide in the South Node.

North Node in Libra in the 7th House/
South Node in Aries in the 1st House

With your South Node in Aries in the 1st house, your gifts of autonomy, individuality, and spontaneity are very much a part of your sense of self; you access them naturally and may not even be consciously aware of them. The 1st house contains everything that we identify as being a fundamental part of who we are. In fact, it's quite difficult to gain perspective on the contents of the 1st house because it's too close to us. Taking control of your life, making your own choices and following your own path, comes so naturally to you that you may not even realize (or care) that these are qualities that are not shared by everyone. The trap of the South Node in Aries in the 1st house is that you continue to simply do your own thing and be entirely focused on expressing your own identity and individuality. Your "take charge" at-

titude can become so second nature to you that you find relating to .
other individuals on a deeper level challenging.

The key to balancing this energy is to work with the North Node
in Libra in the 7th house. The 7th house is the house of one-to-one
relationships, and the biggest challenge of the 7th house is that we
tend to give away the planets in this house, projecting them on others
and experiencing the energies and lessons of these planets through
our relationships. Working with the North Node in Libra may be a
challenge, because you may feel that you lack the qualities of Libra,
and as a result, you will tend to attract people to you who embody
these qualities. These relationships will help you learn how to accept
and integrate the lessons of the North Node in Libra in the 7th house.
With the North Node in Libra in the 7th house, however, experiencing
the North Node lessons through relationships is actually quite appropriate, since your North Node lessons are all about relationships.
These lessons may not always be pleasant, as you attract people into
your life who will set boundaries and limitations on your behavior
and hold you accountable for your actions. However, as you become
more adept at the mechanics of relationships, you will learn how to
find the point of balance and synergy: the point where you are simultaneously expressing your own self and making accommodations for
your partner, and where your partner is expressing himself or herself
while making accommodations for you.

North Node in Libra in the 8th House/
South Node in Aries in the 2nd House

With your South Node in Aries in the 2nd house, your gifts of autonomy, individuality, and spontaneity are important skills and resources
to you. The 2nd house represents our personal resources; anything
that we can call our own belongs in the 2nd house, and everything in
the 2nd house helps reinforce our sense of individuality. Our possessions, resources, skills, and talents, as well as our physical body and
senses, help define and support our identity as an individual. Your
ability to take action, to commit yourself to reaching a goal, and your

drive and energy are the foundation of your ability to generate material wealth and success. Aries' impulsive nature is an indication that you will tend to be reasonably comfortable taking financial risks, and of course you will always be more than willing to pursue any venture whose success depends on your individual ability to perform. The trap of the South Node in Aries in the 2nd house is that you may tend to become too focused on your own resources and abilities, and when something is beyond your ability to accomplish as an individual, this becomes a liability rather than an asset. Aries can be so self-reliant that it never thinks to enlist the help and support of others.

The key to balancing this energy is to work with the North Node in Libra in the 8th house. The 8th house is where we go to find a sense of emotional and soul security and self-worth. We seek a sense of inner peace through letting go of our physical and material attachments and merging with another individual on a deep, healing, and transformational level. The 8th house is also related to shared resources, and, like the 7th house, we often project our 8th-house planets on others, experiencing them through relationships. The North Node in Libra in the 8th house teaches the importance of relating to other individuals and pooling your resources. As always with the Aries/Libra axis, the objective is to maintain your individuality while sustaining a relationship with another individual. Your resources and abundance will increase exponentially when you learn how to combine your individual strengths and gifts with those of other individuals.

North Node in Libra in the 9th House/ South Node in Aries in the 3rd House

With your South Node in Aries in the 3rd house, your gifts of autonomy, individuality, and spontaneity will tend to manifest in your ability to communicate and experience your immediate environment. The 3rd house represents the world that is most familiar to us, and the ways that we make sense of the world through logic, reason, and language. The 3rd house also relates to our individual spiritual practices and beliefs—the ways that we connect to our spirituality outside of the structures of organized religion. Aries energy in the 3rd house often has the feel of the "big fish in the small pond." The strength of Aries' sense of self and its need for individual expression often translates to positions of leadership or prominence within the 3rd-house arena. Aries operates best when it has clearly defined boundaries, and the boundaries of the 3rd house are what is familiar to us. The trap of the South Node in Aries in the 3rd house is to want to *stay* in the 3rd house and never venture into situations that are unfamiliar or that will challenge, expand, or transform your understanding of your daily life and routine. As always, part of the trap of the South Node in Aries involves the desire to become isolated as an individual.

The key to balancing this energy is to work with the North Node in Libra in the 9th house. The 9th house is where we expand our understanding of the world and how we relate to it as an individual, and it includes philosophy, higher education, religion, and long journeys. The North Node in Libra in the 9th house encourages you to explore the unfamiliar territory of the 9th house and, more importantly, to recognize that you are not to make this journey alone. You will learn these new and unfamiliar ideas and concepts from other individuals as you relate to them. You must discover how to balance what you hold to be true for you in your familiar territory with what others hold to be true for them in unfamiliar territory, even as these new discoveries and ideas filter down from the 9th house to the 3rd house and change how you communicate and express your individuality. By working with the North Node in Libra in the 9th house, you will discover how to relate to individuals who perhaps have very different ideals and perspectives from your own; as you maintain balance and harmony in those relationships, the new insights that you discover can help you gain an even better understanding of yourself.

North Node in Libra in the 10th House/
South Node in Aries in the 4th House

With your South Node in Aries in the 4th house, your gifts of autonomy, individuality, and spontaneity are closely related to your private life and family connections. The 4th house represents our home, our families (of choice and of origin), and our connection to our past through our ancestors and tribal heritage. The 4th house is our foundation—the rock on which we build our individual life. It supports us, and through the foundation of the 4th house, we are able to achieve the public, individual accomplishments of the 10th house. You have a very strong sense of your personal connection with your past and with your family, although your natural tendency towards independence and autonomy may mean that you do not maintain a particularly close relationship with your family of origin. The gift of the South Node in Aries in the 4th house is that you are able to act on your own personal needs and find it quite easy to take charge of your personal affairs. The trap of the South Node in Aries in the 4th house is that you may be so good at acting based on your own individuality and your own needs that you sacrifice your personal relationships. Rather than accepting the give and take of family responsibilities, you may tend to go off on your own, either controlling or avoiding your relationships. While you may be accustomed to getting your own way and doing your own thing in your personal life, this type of behavior is far less acceptable outside of your home. When you interact with society, you are bound by society's rules and responsibilities, and you are held accountable for your actions as an individual.

The key to balancing this energy is to work with the North Node in Libra in the 10th house. The 10th house is the most public part of the chart. It relates to our career and our life path; it is where we as individuals find our most significant accomplishments. Your professional relationships are extremely important, and it is through your business partnerships and your professional interactions with society that you will learn how to discover and maintain balance in relationships. In professional relationships, the boundaries are far more obvious than they tend to be in personal relationships. When you enter into a professional agreement or make a commitment through your career or in your public life, you are far more aware of the extent of that commitment and the scope of your responsibilities in fulfilling your part of the agreement. And in the public arena, where you are bound by society's rules, although it is still possible to have things the way that you want them, in order to get them that way, you must learn how to negotiate with others. Your South Node in Aries gifts ensure that you will always maintain your sense of individuality and never sacrifice your needs for the sake of a relationship. But by working with your North Node in Libra in your 10th house you can learn how to experience balanced relationships on a professional level, and then apply these new skills to your personal life.

North Node in Libra in the 11th House/ South Node in Aries in the 5th House

With your South Node in Aries in the 5th house, you encounter your gifts of autonomy, individuality, and spontaneity whenever you explore and express your personal creativity. The 5th house is where we search for security in our identity as an individual. Everything that we can do that makes us feel special and unique, that makes us feel like we deserve the attention and acknowledgment of others, is found in the 5th house—that is why the 5th house is associated with such a mixed bag of concepts including children, the arts, gambling, and love affairs. With the South Node in Aries in the 5th house, you are apt to be drawn to more physical and dynamic expressions of creativity. Aries generally does not have the patience for fine art or painting; Aries is more interested in instant gratification, something that is more frequently found in the performing arts—particularly those that combine physical exertion on the part of the performer with immediate feedback from a live audience. The trap of the South Node in Aries in the 5th house is to become too self-obsessed and too focused on individual expression; at this point the creative process becomes more about self-gratification and less about self-expression.

The key to balancing this energy is to work with the North Node in Libra in the 11th house. The 11th house is where we seek a sense of social and intellectual security. In the 11th house we want to belong to a social community. Within this community, we learn appropriate behavior standards, we participate in group creativity, and we learn how to receive and accept the love of others. With your North Node in Libra in the 11th house, you must learn how to bring your creative energy into your relationships with others, and to discover the joy and importance of experiencing a shared creative experience as opposed to a solitary one. The key here is to join forces with others in a balanced relationship, where each participant brings the full measure of his or her creative energy to the process. Opening up to others in this way and sharing these creative relationships will also help you learn how to receive love and support from others.

North Node in Libra in the 12th House/ South Node in Aries in the 6th House

With your South Node in Aries in the 6th house, you encounter your gifts of autonomy, individuality, and spontaneity in your daily routines. The 6th house contains everything that we must do on a daily basis to maintain our physical existence. The 6th house is related to illness because if we're not taking care of our body with proper nutrition, rest, and exercise on a daily basis, we will get sick. The 6th house is also related to our job and work environment, which is an essential part of maintaining our physical existence. Being active in your daily life is particularly important to you. When this energy is applied to your job environment, you can be perceived as being ambitious and productive, and as an individual who gets things done. As the 6th house is also where we look to serve others, your drive and energy may feel guided by a higher purpose. One of the challenges, of course, is that while your ability to blaze new paths will tend to make you stand out as a leader, more than leadership is needed to manage people. The trap of the South Node in Aries in the 6th house is the tendency to step on toes in the workplace (and in daily life in general); to cross interpersonal boundaries even while trying to be of service. Even though the 6th house relates to being of service to others, Aries energy is not very comfortable with being limited or controlled by others.

The key to balancing this energy is to work with the North Node in Libra in the 12th house. Where the 6th house relates to how we take care of our physical health, the 12th house is how we take care of our spirit and our soul. In the 12th house we encounter our need to feel a part of a spiritual and emotional community; to let go of our individuality and to merge once again with the universe. We retreat to the 12th house whenever we need to take a break from the demands of our daily routines. With the North Node in Libra in your 12th house, your individual spiritual pursuits will help you discover how to bring balance and harmony into your personal relationships. The 12th house is where we go when we want to escape from the physical world for a while, and one way to experience the 12th house is through meditation. As you explore your 12th house, you will begin to gain a more intuitive and instinctive sense of interpersonal boundaries. The North Node in Libra can also help you avoid the ego-oriented traps of the South Node in Aries, which in turn can also help you relate better to other individuals without feeling threatened. Ultimately, you can learn how to bring your spiritual perspective into your daily routine.

4

THE TAURUS/SCORPIO NODAL AXIS

The Taurus/Scorpio nodal axis relates to the universal cycles of birth and death, of growth and destruction. The purpose of this axis is to learn to surrender to the natural cycles of life, and to be open to change when it is necessary. As fixed signs, both Taurus and Scorpio are concerned with sustaining and maintaining, and with the core issue of self-worth. Taurus seeks to sustain and maintain on the physical and material plane, and represents the building side of the axis. Scorpio, on the other hand, seeks to sustain and maintain on the emotional and spiritual plane, which necessitates the tearing down of physical boundaries; and therefore Scorpio represents the destroying side of this axis.

The purpose of Taurus is to sustain and maintain, to stabilize and support the sense of individual identity and initiation that comes from Aries. Taurus reinforces the sense of self through the physical and the material. Taurus defines the self through physical experience and finds validation through the five senses and all forms of interaction with the material plane. Through working with the physical, Taurus seeks to build a structure that will support our sense of individuality and separation from others that we first experienced through

Aries. Taurus, however, can become too focused on the material, the physical, and the practical. The very things that helped support and define our new sense of self can easily become the things that limit and confine us, and hinder our ability to truly grow. Taurus can become attached to the physical and the material, and can lose the ability to distinguish between the true, eternal self and the physical manifestations and extensions of the self.

The purpose of Scorpio on the other hand is to break down these illusions, to strip away all of the concepts that we have used to confine and define our true self, beginning with the physical and ultimately ending only when the ego, the core of our illusions of separation, has been killed, at least temporarily. Scorpio seeks to reinforce our emotional and spiritual self-worth by experiencing deep and transformational connections with others. When we let our ego die, we can merge with another individual on a deep emotional and spiritual level. At least for a moment, we can experience a profound sense of connection to something bigger than ourselves, and be reminded that we are truly all connected and part of all of creation. Scorpio can get carried away in the same way as Taurus, however, and become obsessed with a continual process of destruction and transformation. Scorpio is often tempted to explore all of our hidden fears, each and every one of our buried emotions, and be transformed by this experience. Scorpio doesn't care about the external turmoil that this inner journey tends to create, because for Scorpio, the only true reality exists on the inner, emotional, and spiritual levels.

The Taurus/Scorpio axis relates very strongly to the myth of Persephone. Persephone, the daughter of Demeter and the goddess of the spring, was kidnapped by Hades, the ruler of the underworld. Demeter was so upset that she punished the world: nothing would grow, and for the first time the world experienced fall and winter. Ultimately, Demeter was able to gain the return of her daughter, but as Persephone had eaten six pomegranate seeds while she was in the underworld, she could never truly leave and was bound to Hades. An arrangement was made wherein Persephone would spend six months of the year with Hades in the underworld, and six months of the year with Demeter aboveground. When Persephone is in the upper world, we have spring, the rebirth of life, and everything begins to grow again. When Persephone is in the underworld, we have fall and winter, when things begin to die and the earth is barren until the next spring, when the cycle begins again.

True growth occurs in cycles, and in order for us to be able to continue to grow, a part of us must first die. The Taurus/Scorpio nodal axis teaches us how to get in touch with these cycles; to understand when it is time to put down roots and grow, and when it is time to tear down some of what we have created, so that we can eventually rebuild structures that will allow us to grow even more.

North Node Taurus/South Node Scorpio

With your South Node in Scorpio, you may have experienced great change and transformation in past lifetimes. This experience and familiarity with constant change, with intense emotional transformation, does not necessarily mean that it is something you enjoy or are comfortable with in this lifetime. In fact, the South Node in Scorpio can sometimes indicate an underlying fear of turmoil and transformation. The gifts of the South Node in Scorpio do prepare you for transformation, however. You may not like it in the underworld, but at least you know your way around there. When you encounter transformative emotions and events, when you experience emotional upheaval in your life, you at least can be well prepared.

The challenge of the South Node in Scorpio is that because the constant drama and emotional turmoil of Scorpio is so familiar to you, even though you may not like it, you may take it for granted and assume that there is nothing you can do about it. By working with the energy of your North Node in Taurus, you can begin to rebuild from the ashes. You may continue to experience great transformations in your life, but your lesson involves learning how to maintain an island of calm through all of the upheaval, and committing to rebuild patiently and grow whenever you have the chance. With the South Node in Scorpio and the North Node in Taurus, sometimes Persephone has to be reminded that it's time to come back to the surface so that the world can experience spring again.

North Node in Taurus in the 1st House/
South Node in Scorpio in the 7th House

With your South Node in Scorpio in the 7th house, you will encounter your South Node gifts of transformation, change, and emotional strength through your relationships. One of the challenges of the 7th house, however, is that we tend to give away planets in our 7th house, projecting them on others. You may feel that you lack the Scorpio qualities, and therefore you may tend to attract people into your life who embody them. Until you are able to accept the gifts of the South Node in Scorpio as being a part of you, you will experience and encounter them through your relationships. This can manifest as attracting partners who are exceptionally emotionally intense, whose lives seem to be in a constant state of upheaval, and who will tend to draw you into the trap of the South Node in Scorpio and create havoc and unnecessary crisis and drama in your life. You must first recognize that the drama and emotional intensity in your relationships comes from you, and that the ability to experience these intense, transformational emotional bonds is actually the gift of the South Node in Scorpio. The trap is to believe that you will always *have* to experience crisis and change and will never be able to enjoy stability and peace in your relationships.

Working with your North Node in Taurus in the 1st house can help you balance this energy. However, we face a similar challenge with planets in the 1st house—while we tend to project the 7th house on others, we tend to overlook planets in the 1st house because they are such a fundamental part of who we are as an individual. With the North Node in Taurus in your 1st house, you have the strength, the foundation, and the resources to withstand whatever emotional upheaval the South Node in Scorpio can create. You simply have to cultivate a more conscious awareness of this fact. Your relationships are apt to be quite intense, even when you begin to work consciously with the South Node, because through your relationships you encounter the destructive, transformational energy of Scorpio. By balancing this with an increased awareness of your North Node in Taurus in the 1st house, you can make sure that the Scorpio energy won't completely disrupt your life: you can experience the emotional bonds in your relationships, and then you can use your Taurus North Node energy to rebuild and to grow as an individual.

North Node in Taurus in the 2nd House/
South Node in Scorpio in the 8th House

With your South Node in Scorpio in the 8th house, you will encounter and experience your South Node gifts of transformation, change, and emotional strength when you are experiencing close emotional connections with other individuals. The 8th house is where we go to find a sense of emotional and soul security and self-worth. We seek a sense of inner peace through letting go of our physical and material attachments and merging with another individual on a deep, healing, and transformational level. The 8th house is also related to shared resources, and, like the 7th house, we often project our 8th-house planets on others, experiencing them through relationships. The experience of merging with another individual comes very easily to you; you have a powerful need to be able to share yourself totally with another individual, merging emotions, spirit, soul, and physical resources as well. The trap of the South Node in Scorpio in the 8th house involves pursuing these intense connections in every relationship in your life. Not every individual is willing or able to connect on the levels that Scorpio demands. You must be careful not to lose all sense of interpersonal boundaries, and not to attach your sense of self-worth to your ability to bond with others and to share these intense emotional experiences.

The key to balancing this energy is to work with the North Node in Taurus in the 2nd house. The 2nd house represents our personal resources; anything that we can call our own belongs in the 2nd house, and everything in the 2nd house helps reinforce our sense of individuality. Our possessions, resources, skills, and talents, as well as our physical body and senses, help define and support our identity as an individual. The North Node in Taurus in the 2nd house teaches you that not everything in your life must be shared with other individuals. You must learn how to build your own sense of self-worth, and to create and sustain your own individual resources. You may certainly choose to share these resources with others from time to time, and you are still able to experience the Scorpio emotional bonds and transformations in your relationships. The difference is that when you incorporate the lessons of the North Node in Taurus in the 2nd house, you will maintain your individuality and not become entirely dependent on the emotional energy of others.

North Node in Taurus in the 3rd House/
South Node in Scorpio in the 9th House

With your South Node in Scorpio in the 9th house, you encounter your gifts of transformation, change, and emotional strength on a more philosophical and intellectual level. The 9th house is where we expand our understanding of the world and how we relate to it as an individual, and it includes philosophy, higher education, religion, and long journeys. When the intense, focused energy of Scorpio is applied to 9th-house matters, the result is an almost insatiable curiosity and a drive to understand the core, fundamental, hidden essence of an idea. When it comes to philosophical, religious, or spiritual matters, the South Node in Scorpio in the 9th house will constantly look deeper and deeper, always seeking to understand and connect on the most primal and fundamental level. The trap of the South Node in Scorpio in the 9th house is the tendency to be so focused on discovering what lies beneath the surface that you're never able to actually integrate and apply what you've discovered in your life.

The key to balancing this energy is to work with the North Node in Taurus in the 3rd house. The 3rd house relates to our familiar environment, and to how we communicate, reason, and make connections within that environment. The 3rd house also relates to our individual spiritual practices and beliefs—the ways that we connect to our spirituality outside of the structures of organized religion. The lesson, and the challenge here, is to be able to decide when you've delved deep enough, and to tell the South Node in Scorpio that it's time to stop digging—you need to make use of what you've learned so far and apply it to something tangible and concrete in your life. As always, when we're working with the Taurus/Scorpio axis, the most important lesson is to become attuned to the natural cycle of death and rebirth. The South Node in Scorpio in the 9th house works to tear down our beliefs, our ideas, and our perceptions, and seeks a deeper understanding through philosophy, religion, and higher education. The North Node in Taurus must periodically step in and create new beliefs and perceptions from these new discoveries, and test them out for a while. When their time is up, it's once again time for the South Node in Scorpio in the 9th house to look deeper to define and explore your relationship to the universe.

North Node in Taurus in the 4th House/ South Node in Scorpio in the 10th House

With your South Node in Scorpio in the 10th house, you encounter your South Node gifts of transformation, change, and emotional strength as a part of your public self. The 10th house is where we seek to create a tangible manifestation of our individual identity. It is where we want to make our mark on the world, and where we make our contributions to society as an individual. The 10th house is often associated with our career, but more accurately it is our life path, which is often quite different from our job. Your life path is one that involves a great deal of transformation and change, and you may tend to be acutely aware of every change in your public life, social standing, or career. This is not to say that you will necessarily experience great transformations and changes in your life path, but on some level you may both expect and fear these changes. The trap of the South Node in Scorpio in the 10th house is that you may divert a great deal of energy towards your public and professional life, always being certain that you are able to maintain some sense of control, and always being ready to cope with whatever crisis may come your way. While you certainly have the gift and ability to be able to reinvent yourself time and again, this ability may come at the expense of your personal life.

The key to balancing this energy is to work with the North Node in Taurus in the 4th house. While the 10th house is the most public part of the chart, and also the most focused on individuality, the 4th house is the most private part of the chart, and the area where we are the most connected to others. The 4th house is our home; it is where we connect with our families (both of origin and of choice); but more than this, the 4th house represents our roots, our past, our ancestry, and our tribal heritage. It is the foundation on which we build our individuality and our public self. In a sense, the North Node in Taurus in the 4th house is about allowing yourself to put down roots, to connect with your inner strength. The 4th house is our anchor, our foundation, and this is found both within ourselves, and also through our home and family connections. By allowing yourself to rely on the support and love of your family (either of origin or of choice) and by allowing yourself to trust that your family will remain constant, you will find that you are less at the mercy of the winds of change in your public life, and more able to take full advantage of the opportunities that each new transformation offers you.

North Node in Taurus in the 5th House/
South Node in Scorpio in the 11th House

With your South Node in Scorpio in the 11th house, you encounter your South Node gifts of transformation, change, and emotional strength when you socialize with your friends and peers. The 11th house is where we seek a sense of social and intellectual security. In the 11th house we want to belong to a social community. Within this community, we learn appropriate behavior standards, we participate in group creativity, and we learn how to receive and accept the love of others. Your relationships with your friends are likely to be quite intense. The Scorpio energy in the 11th house often indicates a need and a longing to share a deep emotional bond and connection with our friends. This bond can often be formed when a group of people experience a traumatic situation together, and the emotional connections formed can ultimately be healing for all involved. This, however, can lead to the trap of the South Node in Scorpio in the 11th house, that the only way to experience an emotional bond with others on a sustained basis is to continually experience crisis as a group. The trap of the South Node tells you that if you're not experiencing catharsis in the group, then you're not truly connected with the group; and your sense of self-worth can easily become tied to your feelings of group acceptance.

The key to balancing this energy is to work with the North Node in Taurus in the 5th house. The 5th house is where we search for security in our identity as an individual. Everything that we can do that makes us feel special and unique, that makes us feel like we deserve the attention and acknowledgment of others, is found in the 5th house—that is why the 5th house is associated with such a mixed bag of concepts including children, the arts, gambling, and love affairs. With your North Node in Taurus in the 5th house, you must learn to express your personal creativity in a very tangible manner. Taurus energy is quite creative and artistic by nature, and the North Node in Taurus in the 5th house points out the importance of having a constructive, creative outlet for your self-expression. Exploring your individual creative outlets will help you maintain a more healthy sense of self-worth, one that is not so dependent on sharing in the emotional and traumatic experiences of others.

North Node in Taurus in the 6th House/
South Node in Scorpio in the 12th House

With your South Node in Scorpio in the 12th house, you may not be consciously aware of your gifts of transformation, change, and emotional strength, even though they are very evident to other people. The 12th house is like our shadow: we can't see it directly because we're usually facing the light (of consciousness, that is), but it's very visible to everyone else. Our awareness of our 12th house seems to come from our unconscious and our subconscious. You have an unconscious and intuitive need for change and transformation; the accompanying fear of change and transformation, however, is likely to be more obvious to you. Because your need for change and for deep emotional bonds is largely unconscious, you may not be aware that you actually create the various crises and upsets in your life yourself. The trap of the South Node in Scorpio in the 12th house is the tendency to buy into your fear of change and your perception that you are not capable of building any kind of stability in your life because something always comes along to disrupt things.

The key to balancing this energy is to work with the North Node in Taurus in the 6th house. While the 12th house relates to our emotional and soul needs, the 6th house relates to our routine physical needs. The 6th house is related to illness because if we're not taking care of our body with proper nutrition, rest, and exercise on a daily basis, we will get sick. The 6th house is also related to our job and work environment, which is an essential part of maintaining our physical existence. The North Node in Taurus in the 6th house emphasizes that every day you must learn to focus on the things that you are creating. The process of building up and tearing down is ongoing and eternal. As you begin to integrate the Taurus North Node energy, you can begin to release your fear of change. Everything in your life has its own cycle of creation and destruction; so long as you don't become attached, you won't be upset when it's gone. Creating and maintaining a daily routine is essential; the daily touchstones of your life are the things that will provide you with tangible reassurance and allow you to experience the change and transformations in your life with a minimum of fear and discomfort.

North Node in Taurus in the 7th House/
South Node in Scorpio in the 1st House

With your South Node in Scorpio in the 1st house, your gifts of transformation, change, and emotional strength are very much a part of your sense of self; you access them naturally and may not even be consciously aware of them. The 1st house contains everything that we identify as being a fundamental part of who we are. In fact, it's quite difficult to gain perspective on the contents of the 1st house because it's too close to us. You may tend to define yourself in terms of the deep and powerful emotional connections that you form with others. The trap of the South Node in Scorpio in the 1st house is that Scorpio's fundamental need to tear down the ego and to experience death and rebirth, change, and transformation can make it difficult for you to discover and express your true self because you're always in the process of tearing it down, of looking even deeper, searching for the most hidden, most spiritual aspects. This can lead to a sense of isolation, both because the South Node in Scorpio in the 1st house tends to be introspective and because the persona that you project to the world is so focused and intense that you may intimidate people who aren't ready to bear their souls with you.

The key to balancing this energy is to work with the North Node in Taurus in the 7th house. The 7th house is the house of one-to-one relationships, and the biggest challenge of the 7th house is that we tend to give away the planets in this house, projecting them on others and experiencing the energies and lessons of these planets through our relationships. Working with the North Node in Taurus may be a challenge, because you may feel that you lack the qualities of Taurus. As a result, you will tend to attract people to you who embody these qualities. These relationships will help you learn how to accept and integrate the lessons of the North Node in Taurus in the 7th house. You may find that you tend to attract and be attracted to stable, practical, grounded individuals; individuals, in other words, who embody the Taurus energy of your North Node. Rather than falling into the trap of the South Node in Scorpio in the 1st house and allowing these individuals to become your anchor, letting them contain all of the emotional storms that you experience, these relationships are meant to teach you to discover the core sense of self-worth and stability that you carry within you. You can learn to contain and control your own growth and transformation process. You will generally need to be reminded by others that it's time to stop processing and to take a break, but ultimately you can learn to make the journey back up from the underworld on your own.

North Node in Taurus in the 8th House/
South Node in Scorpio in the 2nd House

With your South Node in Scorpio in the 2nd house, your gifts of transformation, change, and emotional strength are important skills and resources to you. The 2nd house represents our personal resources; anything that we can call our own belongs in the 2nd house, and everything in the 2nd house helps reinforce our sense of individuality. Our possessions, resources, skills, and talents, as well as our physical body and senses, help define and support our identity as an individual. Your gifts of transformation and change are most easily applied to your personal resources. Granted, at times, the South Node in the 2nd house may feel like anything but a gift, particularly when it results in a periodic reversal of fortune. The gift of the Scorpio energy, however, is that you have the courage and the strength to learn from your setbacks and to improve. The trap of the South Node in Scorpio in the 2nd house is that you may come to expect these changes in your resources and to dwell on the pain associated with each loss. Your self-esteem will begin to suffer if you believe that these changes occur because you are not worthy as an individual; that is, if you were good enough, you would be able to experience stability in your life.

The key to balancing this energy is to work with the North Node in Taurus in the 8th house. The 8th house is where we go to find a sense of emotional and soul security and self-worth. We seek a sense of inner peace through letting go of our physical and material attachments and merging with another individual on a deep, healing, and transformational level. The 8th house is also related to shared resources, and, like the 7th house, we often project our 8th-house planets on others, experiencing them through relationships. The lesson of the North Node in Taurus in the 8th house is to allow yourself to trust a partner, to learn how to share your resources with another, and how to accept the support and assistance of a partner in return. Through relationships, and through allowing yourself to trust in the stability of another, you can discover how to build a more stable foundation, how to put down some roots and enjoy a period of security and stability. You may still find that your personal resources tend to fluctuate, but during these times you will be able to draw on the resources of a partner for support, and not have to start over again from scratch.

North Node in Taurus in the 9th House/
South Node in Scorpio in the 3rd House

With your South Node in Scorpio in the 3rd house, your gifts of transformation, change, and emotional strength will tend to manifest in your ability to communicate and experience your immediate environment. The 3rd house represents the world that is most familiar to us, and the ways that we make sense of the world through logic, reason, and language. The 3rd house also relates to our individual spiritual practices and beliefs—the ways that we connect to our spirituality outside of the structures of organized religion. Of course, you may not always recognize these Scorpio experiences as gifts—the constant changes, emotional upheavals, and crises can be very disconcerting, particularly as they relate to our early childhood experiences and early schooling, which are also part of the domain of the 3rd house. The gift of the South Node in Scorpio in the 3rd house is the ability to survive this constant state of flux; to grow stronger and more capable as you survive each transformation. The trap of the South Node in Scorpio in the 3rd house is to get caught up in the daily drama of your life, to buy into the illusion that you will never be able to experience stability in your environment. When you lose sight of the fact that the changes brought about by the South Node in Scorpio in the 3rd house are teaching you how to grow and how to be stronger, you can become despondent. You must learn to keep things in their proper perspective.

The key to balancing this energy is to work with the North Node in Taurus in the 9th house. The 9th house is where we expand our understanding of the world and how we relate to it as an individual, and it includes philosophy, higher education, religion, and long journeys. Part of the challenge of your North Node in Taurus in the 9th house is to be able to find some sense of tangible security through intellectual and spiritual pursuits (which, by their nature, are intangible). The ceremony and routine of organized religion can be a great solace. These touchstones will not change; they have weathered storms far greater than the ones that occur in your sphere of reality, and they still survive. Studying philosophy can also be quite helpful as it can provide you with a seemingly external point of view, which you can then apply to the events in your immediate environment.

North Node in Taurus in the 10th House/
South Node in Scorpio in the 4th House

With your South Node in Scorpio in the 4th house, your gifts of transformation, change, and emotional strength are closely related to your private life and family connections. The 4th house represents our home, our families (of choice and of origin), and our connection to our past through our ancestors and tribal heritage. The 4th house is our foundation—the rock on which we build our individual life. It supports us, and through the foundation of the 4th house we are able to achieve the public, individual accomplishments of the 10th house. The gifts of the South Node in Scorpio in the 4th house may not always feel like they are gifts, because part of what they create is a continually transforming foundation, which can feel quite unsettling. These gifts, however, are likely to motivate you to connect with your past, and form deep emotional bonds with your family. The South Node in Scorpio in the 4th house seeks deep, powerful emotional connections on a soul level; it seeks to support your sense of emotional self-worth by discovering and merging with your tribal roots. The trap of the South Node in Scorpio in the 4th house is to become too obsessed with the past, and too heavily interconnected with the family myth and karma. Your family and your history and heritage are very important, but they do not define you as an individual. You must build on the foundation of your past and express your individuality.

The key to balancing this energy is to work with the North Node in Taurus in the 10th house. The 10th house is the most public part of the chart. It relates to our career and our life path; it is where we as individuals find our most significant accomplishments. With the North Node in Taurus in the 10th house, you are encouraged to quite literally build your own public identity. You must draw on your South Node in Scorpio in the 4th, and the deeper the roots go, the stronger your creation will be. This energy has very much the feeling of the self-made (wo)man: your life path involves building and defining your individual identity as an extension and a unique expression of your family and ancestral connections. Taurus' drive to create things of lasting beauty encourages you to create a life for yourself that makes a valuable contribution to and a lasting impression on society, which in turn will also contribute to the richness and strength of your family.

North Node in Taurus in the 11th House/
South Node in Scorpio in the 5th House

With your South Node in Scorpio in the 5th house, you encounter your gifts of transformation, change, and emotional strength whenever you explore and express your personal creativity. The 5th house is where we search for security in our identity as an individual. Everything that we can do that makes us feel special and unique, that makes us feel like we deserve the attention and acknowledgment of others, is found in the 5th house—that is why the 5th house is associated with such a mixed bag of concepts including children, the arts, gambling, and love affairs. One of the more obvious connections here is between the sexual side of Scorpio and the romantic and sexual nature of the 5th house. The South Node in Scorpio in the 5th house can make passionate love affairs with deep, intense emotional connections particularly appealing to you. The trap of the South Node in Scorpio in the

5th house is to become too focused on your romantic and sexual life, which in turn can cause you to base your sense of self-worth on your ability to experience emotional connections with romantic partners. Sex is far from the only way to make use of this intense creative energy, but often the allure of the South Node in Scorpio in the 5th house is quite strong. In order to find alternative expressions of this energy and applications for your South Node gifts, you must learn to work with the North Node in Taurus in the 11th house.

In order to find alternative expressions of this energy and applications for your South Node gifts, you must learn to work with the North Node in Taurus in the 11th house. The 11th house is where we seek a sense of social and intellectual security. In the 11th house we want to belong to a social community. Within this community, we learn appropriate behavior standards, we participate in group creativity, and we learn how to receive and accept the love of others. With your North Node in Taurus in the 11th house, you must learn to direct your attention and energy towards your friendships in order to balance out and stabilize the passionate turmoil of your romantic relationships. Your friends and peers can help keep you grounded and real. They can help you contain the storms of the Scorpio need for emotional and soul connections, and they can provide you with a way to experience a sense of unity with others through shared creativity that does not require the tearing down of the ego and the emotional and psychological upheavals that Scorpio often requires. Participating in group activities, particularly ones that are creative in nature and that work towards a tangible goal, can help you build a healthier sense of self-worth and avoid the traps of the South Node in Scorpio in the 5th house.

North Node in Taurus in the 12th House/
South Node in Scorpio in the 6th House

With your South Node in Scorpio in the 6th house, you encounter your gifts of transformation, change, and emotional strength in your daily routines. The 6th house contains everything that we must do on a daily basis to maintain our physical existence. The 6th house is re-lated to illness because if we're not taking care of our body with proper nutrition, rest, and exercise on a daily basis, we will get sick. The 6th house is also related to our job and work environment, which is an essential part of maintaining our physical existence. This can present a situation where you might wish that the "gifts" of the South Node in Scorpio could be returned and exchanged for something else. Your relationships with your co-workers, and your work environment in general, are apt to be very intense and emotionally charged. You may find that you are drawn to jobs that involve a great deal of pres-sure and that require the ability to manage emotional crises. As the 6th house is also the place where we look to be of service to others, this energy often relates to health care as well as to psychology and psychiatry. The trap of the South Node in Scorpio in the 6th house is the temptation to become too personally involved in the trauma and drama in your daily life. While part of the gift of the Scorpio South Node is the ability to heal and transform, to help others through their most traumatic experiences, you must learn how to do this without being drawn into these experiences yourself.

The key to balancing this energy is to work with the North Node in Taurus in the 12th house. Where the 6th house relates to how we take care of our physical health, the 12th house is how we take care of our spirit and our soul. In the 12th house we encounter our need to feel a part of a spiritual and emotional community; to let go of our in-dividuality and to merge once again with the universe. We retreat to the 12th house whenever we need to take a break from the demands of our daily routines. With the North Node in Taurus in your 12th house, this retreat becomes a place of great comfort and protection. Spiritual practices such as meditation and yoga will help you sustain your peace of mind. Even your hobbies can have this effect, particularly if they are largely solitary pursuits that involve working with tangible, material things, such as gardening or pottery. By working with your North Node in Taurus in the 12th house and becoming more in tune with your spiritual stability and peace, you will find that you will become more effective when experiencing the Scorpio events in your life; you are able to remain more calm and detached, and are therefore even more capable of helping both yourself and others through times of crisis.

North Node Scorpio/South Node Taurus

Individuals with the South Node in Taurus bring the gifts of stability, creativity, being grounded, and of growth into this lifetime. They have learned well how to create a solid foundation, and how to build on that foundation slowly and steadily to create an integrated sense of self. The trap of the South Node in Taurus, however, is that these individuals may be quite addicted to this illusion of structure and extremely resistant to change. In fact, the idea of change may positively terrify them. Change, of course, is exactly what is required for these individuals. They have taken the Taurus energy as far as they can, and it's time to let some of it die to clear the way for new growth. The North Node in Scorpio tells us that it's time for Persephone to return to the underworld, and it's time for these individuals to let certain parts of themselves die so that other parts can continue to grow.

The South Node in Taurus indicates that you have come from a point of stability and constant growth, that in past lives you have built a very grounded and stable sense of self. In this life, however, you must learn how to let go of that stability and to experience intense transformation and change.

Your lesson involves being able to hold lightly onto your values and possessions, and to make sure that you do not begin to identify with them. You have the gift of being able to find the calm in the storm, of being able to ground and stabilize and bring things back to a practical level. It's time to learn when to let things go, when to give up the illusion of control and to release your attachment to the structures in your life so that you can grow on an emotional and spiritual level.

North Node in Scorpio in the 1st House/
South Node in Taurus in the 7th House

With your South Node in Taurus in the 7th house, you will encounter your South Node gifts of creativity, stability, loyalty, and stamina through your relationships. One of the challenges of the 7th house, however, is that we tend to give away planets in our 7th house, projecting them on others. You may feel that you lack the Taurus qualities, and therefore you may tend to attract people into your life who embody them. Until you are able to accept the gifts of the South Node in Taurus as being a part of you, you will experience and encounter them through your relationships. Rather than acknowledging your South Node in Taurus gifts as your own, you may feel that you lack the ability to be stable and grounded, and instead attract people into your life who embody the stability and security of Taurus. Your lesson is to recognize these qualities in yourself; to realize that you are quite capable of being grounded and practical. The trap of the South Node in Taurus in the 7th house is to always look to other people for stability and security, and to base your sense of self-worth on the opinions of others. The trap of the South Node in Taurus can also lead you to become overly concerned with appearances and material concerns, and to use these as a measure of your own worth and, by projection, of the worth of others.

Working with your North Node in Scorpio in the 1st house can help you balance this energy. However, we face a similar challenge with planets in the 1st house—while we tend to project the 7th house on others, we tend to overlook planets in the 1st house because they are such a fundamental part of who we are as an individual. With the North Node in Scorpio in your 1st house, you have the ability to continually reinvent yourself, to let go of the illusions of material security and to search for a greater truth and spiritual understanding by exploring your emotional and soul nature. Your relationships and your partners will certainly be an integral part of this experience, because they can help to remind you that you do indeed have a strong, solid foundation to work from; however, the journey that you must take is one that must be experienced alone. You must be willing to let go of the security of your relationships, and risk transforming the nature of your relationships as a result of your own journey into your unconscious.

North Node in Scorpio in the 2nd House/
South Node in Taurus in the 8th House

With your South Node in Taurus in the 8th house, you will encounter and experience your South Node gifts of creativity, stability, loyalty, and stamina when you are experiencing close emotional connections with other individuals. The 8th house is where we go to find a sense of emotional and soul security and self-worth. We seek a sense of inner peace through letting go of our physical and material attachments and merging with another individual on a deep, healing, and transformational level. The 8th house is also related to shared resources, and, like the 7th house, we often project our 8th-house planets on others, experiencing them through relationships. You may find that you derive support and comfort from the resources that you share with others. While your South Node gifts make you by nature extremely reliable and rather conservative when it comes to managing and maintaining shared resources, the trap of the South Node in Taurus in the 8th house is that you may become entirely dependent on these resources. Your sense of self-worth can become closely tied to other people's resources, and since you ultimately have no control over these resources, you have the potential to give away much of your personal power in your relationships. You must learn to release your attachments to these resources, and ultimately to release your attachment to the physical plane as a security blanket.

The key to balancing this energy is to work with the North Node in Scorpio in the 2nd house. The 2nd house represents our personal resources; anything that we can call our own belongs in the 2nd house, and everything in the 2nd house helps reinforce our sense of individuality. Our possessions, resources, skills, and talents, as well as our physical body and senses, help define and support our identity as an individual. The North Node in Scorpio in the 2nd house challenges you to take risks, to leave behind the constricting security of the 8thhouse shared resources, and to explore the deeper levels of your own gifts and talents. Part of the lesson of the North Node in Scorpio in the 2nd house is to learn that one of the laws of abundance dictates that we must not hold onto things; that in order to experience true abundance in our life, we must be willing to let that energy flow, and that means experiencing periodic changes of fortune. With the South Node in Taurus in the 8th house, you may find this idea quite threatening. You must learn to embrace the truth of the North Node in Scorpio in the 2nd house and know that even when your material resources disappear, that this is simply a natural part of the cycle of abundance and growth, and that you will always receive more.

North Node in Scorpio in the 3rd House/ South Node in Taurus in the 9th House

With your South Node in Taurus in the 9th house, you encounter your gifts of creativity, stability, loyalty, and stamina on a more philosophical and intellectual level. The 9th house is where we expand our understanding of the world and how we relate to it as an individual, and it includes philosophy, higher education, religion, and long journeys. You can find great solace and security in these pursuits; your beliefs and ideas can become almost tangible, and you take great comfort in the fact that much of this knowledge has endured for centuries without change. The trap of the South Node in Taurus in the 9th house, however, is the tendency to become too attached to these ideas; to be afraid to question them or even to put them to practical use. This is the point where spirituality and religion become dogma; where ideas are sustained and reinforced because they always have been. And the more you fall into the trap of the South Node in Taurus in the 9th house, the more you cling to and defend these ideas, the less comfort and security they provide to you.

The key to balancing this energy is to work with the North Node in Scorpio in the 3rd house. The 3rd house relates to our familiar environment, and to how we communicate, reason, and make connections within that environment. The 3rd house also relates to our individual spiritual practices and beliefs—the ways that we connect to our spirituality outside of the structures of organized religion. The North Node in Scorpio in the 3rd house urges you to look beyond the dogma and the abstract philosophies and to seek a deeper meaning, a deeper understanding of the core concepts. More importantly, it urges you to look for this understanding by applying these ideas and concepts in your daily life and your immediate worldview. It's not enough to pursue your spiritual connections by attending religious services; you must discover how these connections apply to every aspect of your world. As you explore and investigate, many of your closely held beliefs are going to be forced to change, and this, of course, is very threatening to the South Node in Taurus. But part of the lesson of the North Node in Scorpio in the 3rd house is to accept that even beliefs have their own cycles of creation and destruction; we must learn to let them go when they have outlived their usefulness to us.

North Node in Scorpio in the 4th House/
South Node in Taurus in the 10th House

With your South Node in Taurus in the 10th house, you encounter your South Node gifts of creativity, stability, loyalty, and stamina as a part of your public self. The 10th house is where we seek to create a tangible manifestation of our individual identity. It is where we want to make our mark on the world, and where we make our contributions to society as an individual. The 10th house is often associated with our career, but more accurately it is our life path, which is often quite different from our job. Building a stable, structured, grounded, and lucrative professional life is something that can come quite easily to you. However, you must also be aware that the traps of the South Node in Taurus in the 10th house involve becoming too concerned with your public face, with your career, and with your social standing. You can become too attached to the material trappings of success and begin to identify too strongly with your ambitions and accomplishments.

The key to balancing this energy is to work with the North Node in Scorpio in the 4th house. While the 10th house is the most public part of the chart, and also the most focused on individuality, the 4th house is the most private part of the chart, and the area where we are the most connected to others. The 4th house is our home; it is where we connect with our families (both of origin and of choice); but more than this, the 4th house represents our roots, our past, our ancestry, and our tribal heritage. It is the foundation on which we build our individuality and our public self. With your North Node in Scorpio in your 4th house, you must be willing to explore the myths of your past and of your family, to tear down the old structures and stories and to dig deep in order to discover your own personal connection to your family and to your tribe. You can carry with you your South Node gifts of patience, strength, and stability. You realize that the changes you must initiate are all a part of the natural cycle, and that as you discover your true roots, you will once again be able to concentrate on growing.

North Node in Scorpio in the 5th House/
South Node in Taurus in the 11th House

With your South Node in Taurus in the 11th house, you encounter your South Node gifts of creativity, stability, loyalty, and stamina when you socialize with your friends and peers. The 11th house is where we seek a sense of social and intellectual security. In the 11th house we want to belong to a social community. Within this community, we learn appropriate behavior standards, we participate in group creativity, and we learn how to receive and accept the love of others. Your friendships are apt to be a tremendous source of strength and support for you. You may, in fact, have a number of lifelong friends with whom you keep in touch even though your lives may have drifted apart and you no longer share much common ground other than your past. The trap of the South Node in Taurus in the 11th house is to become too dependent on your friendships; when you become attached to these relationships, your sense of self-worth can become involved so that it is entirely bound up in your ability to feel loved and accepted by your peers.

The key to balancing this energy is to work with the North Node in Scorpio in the 5th house. The 5th house is where we search for security in our identity as an individual. Everything that we can do that makes us feel special and unique, that makes us feel like we deserve the attention and acknowledgment of others, is found in the 5th house—that is why the 5th house is associated with such a mixed bag of concepts including children, the arts, gambling, and love affairs. With your North Node in Scorpio in the 5th house, you are encouraged to let go of your group associations for a while and to spend time exploring and discovering exactly what makes you special, unique, and worthy of being loved. You are apt to approach all 5th-house activities with a great deal of passion and intensity, and this most certainly includes your romantic involvements. Surrendering to the Scorpio North Node means being willing to grow and change as an individual. You can also learn that it's okay to let go of old friendships that are entirely based on past experiences. Your important friendships exist and thrive because of who you are now, not because of who you once were.

North Node in Scorpio in the 6th House/
South Node in Taurus in the 12th House

With your South Node in Taurus in the 12th house, you may not be consciously aware of your gifts of creativity, stability, loyalty, and stamina, even though they are very evident to other people. The 12th house is like our shadow: we can't see it directly because we're usually facing the light (of consciousness, that is), but it's very visible to everyone else. Our awareness of our 12th house seems to come from our unconscious and our subconscious. You have an unconscious need for material security and stability. The 12th house relates to our spiritual connections, and the gifts of the South Node in Taurus in the 12th house are the ability to understand and truly appreciate that we are eternal beings, and that nothing true or real can ever be destroyed. The trap of the South Node in Taurus in the 12th house, however, is an unconscious fear of change; an attachment to the physical and to appearances.

The key to balancing this energy is to work with the North Node in Scorpio in the 6th house. While the 12th house relates to our emotional and soul needs, the 6th house relates to our routine physical needs. The 6th house is related to illness because if we're not taking care of our body with proper nutrition, rest, and exercise on a daily basis, we will get sick. The 6th house is also related to our job and work environment, which is an essential part of maintaining our physical existence. The North Node in Scorpio in the 6th house teaches you that you must be willing to explore beneath the surface of everything in your life; you must never become attached to the physical or the material because that is only illusion—the greater spiritual truths lie far beneath the surface. Ultimately, the goal is to learn to draw on your spiritual knowledge that we are eternal, to take comfort and support from that, and then to be able to embrace change and transformation when it occurs in your life (as it most certainly will). Much of this lesson may tend to play out in your work environment and in your relationships with your co-workers, although it can also show up through health issues as well. The more you try to resist the changes, the more difficult they will be for you. Your lessons are to trust and to surrender to the natural cycles of the universe.

North Node in Scorpio in the 7th House/
South Node in Taurus in the 1st House

With your South Node in Taurus in the 1st house, your gifts of creativity, stability, loyalty, and stamina are very much a part of your sense of self; you access them naturally and may not even be consciously aware of them. The 1st house contains everything that we identify as being a fundamental part of who we are. In fact, it's quite difficult to gain perspective on the contents of the 1st house because it's too close to us. Your creativity, your constancy, your groundedness, and your strength all come so naturally to you that you may not even realize (or care) that these are qualities that are not shared by everyone.

The trap of the South Node in Taurus in the 1st house, however, is to become too addicted to your perception of yourself; to invest too much in your constancy and your own personal reality and to become rigid, stubborn, and ultimately terrified of change. The more you resist change, of course, the more difficult it will be. Change is inevitable, and with your North Node in Scorpio in your 7th house, it's going to come as a result of your relationships.

The key to balancing this energy is to work with the North Node in Scorpio in the 7th house. The 7th house is the house of one-to-one relationships, and the biggest challenge of the 7th house is that we tend to give away the planets in this house, projecting them on others and experiencing the energies and lessons of these planets through our relationships. Working with the North Node in Scorpio may be a challenge, because you may feel that you lack the qualities of Scorpio, and as a result, you will tend to attract people to you who embody these qualities. These relationships will help you learn how to accept and integrate the lessons of the North Node in Scorpio in the 7th house. You may find that you both attract and are on some level attracted to people who value their emotional connections above their material possessions, and whose lives seem fraught with change, crisis, passion, intensity, and transformation. Your relationships with these individuals will force you to reevaluate your own sense of self and your own identity; the more you resist, the more you condemn others for their instability and their recklessness, the more drastic the impact they will have on you and the more pressure you will feel to change yourself. If you are able to surrender to these relationships and embrace the lessons of the North Node in Scorpio in the 7th house, you can allow your sense of self to dissolve as you merge with another individual on a primal, emotional, and soul level. When you are reborn after this experience, you can rebuild your sense of self—and discover, to the great joy of your South Node in Taurus, that you are actually able to grow as a result of the experience.

North Node in Scorpio in the 8th House/ South Node in Taurus in the 2nd House

With your South Node in Taurus in the 2nd house, your gifts of creativity, stability, loyalty, and stamina are important skills and resources to you. The 2nd house represents our personal resources; anything that we can call our own belongs in the 2nd house, and everything in the 2nd house helps reinforce our sense of individuality. Our possessions, resources, skills, and talents, as well as our physical body and senses, help define and support our identity as an individual. You are likely to take a very conservative approach to your resources, and to derive a great deal of pleasure and comfort from your material wealth and physical abundance. The trap of the South Node in Taurus in the 2nd house, however, is to become attached to the material plane and to tie your sense of self-worth and of value as an individual to your possessions. The more empty you feel, the more you will be driven to acquire; and the more you acquire, the emptier you will feel.

The key to balancing this energy is to work with the North Node in Scorpio in the 8th house. The 8th house is where we go to find a sense of emotional and soul security and self-worth. We seek a sense of inner peace through letting go of our physical and material attachments and merging with another individual on a deep, healing, and transformational level. The 8th house is also related to shared resources, and, like the 7th house, we often project our 8th-house planets on others, experiencing them through relationships. The North Node in Scorpio in the 8th house teaches that you must be willing to let go of your attachment to the physical and to look for validation and support of your sense of self-worth through emotional connections with other individuals. Combining your resources with a partner is one way of giving up some of your control over your resources. Ultimately, the North Node in Scorpio in the 8th house can teach you how to meet your emotional and security needs directly through relationships, rather than indirectly by pursuing comfort on the physical plane.

North Node in Scorpio in the 9th House/
South Node in Taurus in the 3rd House

With your South Node in Taurus in the 3rd house, your gifts of creativity, stability, loyalty, and stamina will tend to manifest in your ability to communicate and experience your immediate environment. The 3rd house represents the world that is most familiar to us, and the ways that we make sense of the world through logic, reason, and language. The 3rd house also relates to our individual spiritual practices and beliefs—the ways that we connect to our spirituality outside of the structures of organized religion. With the South Node in Taurus in the 3rd house, what you may be holding onto has more to do with your comfort with your immediate surroundings, with what is familiar to you. The 3rd house also relates to our early education, and you may find that you tend to hold on to the ideas and concepts that you learned at a young age. You understand how this world works, it's familiar and comfortable to you, and you know exactly how you fit in. The trap of the South Node in Taurus in the 3rd house, of course, is the desire to want to stay within the boundaries of this familiar world; to always stick with what is familiar and expected; and to never grow, to never venture out of your neighborhood either literally or figuratively.

The key to balancing this energy is to work with the North Node in Scorpio in the 9th house. The 9th house is where we expand our understanding of the world and how we relate to it as an individual, and it includes philosophy, higher education, religion, and long journeys. The North Node in Scorpio in the 9th house indicates that you can have your sense of self, your sense of emotional connection and worth, and your understanding of how you fit into the world completely transformed once you begin to explore the unfamiliar and to investigate and immerse yourself in different philosophies, religions, cultures, and schools of thought. The way for you to experience this is through total immersion, through a deep emotional connection—in other words, through Scorpio. The key here is to learn how to be able to let go of your familiar perceptions of the world long enough to immerse yourself in the new and the unfamiliar and to be transformed by it. Then you can incorporate this new knowledge, this new understanding of yourself and of the world, into yourself so that it's no longer strange and unfamiliar. You can learn how to expand your experience of the 3rd house by exploring and bringing back experiences from the 9th house.

North Node in Scorpio in the 10th House/
South Node in Taurus in the 4th House

With your South Node in Taurus in the 4th house, your gifts of creativity, stability, loyalty, and stamina are closely related to your private life and family connections. The 4th house represents our home, our families (of choice and of origin), and our connection to our past through our ancestors and tribal heritage. The 4th house is our foundation—the rock on which we build our individual life. It supports us, and through the foundation of the 4th house, we are able to achieve the public, individual accomplishments of the 10th house. You draw a great deal of security from your roots and your family, and the traditions of your family are likely to be very important to you. While the gift of the South Node in Taurus in the 4th house is your ability to draw strength and sustenance from your family and your past, the trap of the South Node in Taurus in the 4th house is to become so dependent on this support and security that you never break away and create your own individual identity.

The key to balancing this energy is to work with the North Node in Scorpio in the 10th house. The 10th house is the most public part of the chart. It relates to our career and our life path; it is where we as individuals find our most significant accomplishments. The Scorpio energy of change, transformation, death and rebirth is closely linked with your life path, and as you set out into the world, you may find that you are drawn to careers that involve discovery, research, change, and transformation—of yourself, others, and society as a whole. The North Node in Scorpio journey can often feel like it takes you far from the security and comfort of your home and family; what you must realize is that you can always draw on that strength and support. The discoveries that you make, the emotions that you experience, these will all help define you as an individual and in turn allow you to re-connect with your family and your roots. You may not relate in the same way or on the same level as you did before, but your new con-nections to your family will be stronger, and you will be contributing your individuality to the family energy.

North Node in Scorpio in the 11th House/
South Node in Taurus in the 5th House

With your South Node in Taurus in the 5th house, you encounter your gifts of creativity, stability, loyalty, and stamina whenever you ex-plore and express your personal creativity. The 5th house is where we search for security in our identity as an individual. Everything that we can do that makes us feel special and unique, that makes us feel like we deserve the attention and acknowledgment of others, is found in the 5th house—that is why the 5th house is associated with such a mixed bag of concepts including children, the arts, gambling, and love affairs. This placement can indicate a great deal of natural talent in the arts, and the ability to truly connect with your soul essence through the act of creating. The creative process can be a solitary one, however, and the desire to create, to produce tangible expressions of yourself, can also be connected with a fear that if you turn your focus outward and begin to experience yourself as an equal member of the group, that this will in some way change or diminish your sense of self. This is the trap of the South Node in Taurus in the 5th house.

The key to balancing this energy is to work with the North Node in Scorpio in the 11th house. The 11th house is where we seek a sense of social and intellectual security. In the 11th house we want to belong to a social community. Within this community, we learn appropriate behavior standards, we participate in group creativity, and we learn how to receive and accept the love of others. The North Node in Scor-pio in the 11th house seeks a fundamental emotional connection that allows the ego to die and then be reborn with an increased sense of self. You can encounter this transformation and growth when you are involved in activities with your peers, with groups of individuals of your own choosing who share common interests and who each have something unique and personally creative to share. Although the South Node in Taurus can be very resistant to change, by connecting with your South Node gifts, your personal creativity, and then contributing this energy to accomplish something greater through a group effort, yes, you may temporarily give up personal acknowledgment; but what you will gain is a greater and even more expanded ability to create, a new and more comprehensive understanding of how to work with and take advantage of your South Node gifts.

North Node in Scorpio in the 12th House/
South Node in Taurus in the 6th House

With your South Node in Taurus in the 6th house, you encounter your gifts of creativity, stability, loyalty, and stamina in your daily routines. The 6th house contains everything that we must do on a daily basis to maintain our physical existence. The 6th house is related to illness because if we're not taking care of our body with proper nutrition, rest, and exercise on a daily basis, we will get sick. The 6th house is also related to our job and work environment, which is an essential part of maintaining our physical existence. With the South Node in Taurus in the 6th house, however, the operative word is *routine*. You are likely to take great comfort and find security in your daily routines and rituals, and you may find any disruptions or potential disruptions to these routines to be very threatening and unsettling. The focus of the South Node in Taurus in the 6th house is to build and create a stable, lasting physical existence. You will be most comfortable in work environments that are conducive to predictable and routine tasks; you are certainly not looking for excitement in your daily life. The more attached to your routines you become, however, the more you will fall into the trap of the South Node in Taurus in the 6th house. As you become too attached to your routines and to the physical trappings of your daily life, you find that you are devoting more and more energy towards trying to prevent any changes or disruptions to these routines;

but because physical existence is transitory by nature, you're always fighting a losing battle.

The key to balancing this energy is to work with the North Node in Scorpio in the 12th house. Where the 6th house relates to how we take care of our physical health, the 12th house is how we take care of our spirit and our soul. In the 12th house we encounter our need to feel a part of a spiritual and emotional community; to let go of our individuality and to merge once again with the universe. We retreat to the 12th house whenever we need to take a break from the demands of our daily routines. With the North Node in Scorpio in the 12th house, what your time spent in spiritual pursuits is trying to teach you is that what is eternal and unchanging is our soul self, and that we can freely release our attachments to the physical and material plane. The Scorpio energy encourages you to look deep within yourself to rediscover the truth that we all know, and yet that we so easily forget. The physical reality that we experience is only an illusion. We must respect it and we are certainly able to enjoy it for what it is, but we must also recognize that it will change and be transformed, and that these changes and transformations in no way affect the true essence of our existence.

5

THE GEMINI/SAGITTARIUS NODAL AXIS

The Gemini/Sagittarius nodal axis is the axis of the mind. The purpose of this axis is to learn how to balance the lower mind with the higher mind; the immediate environment with the cosmos; knowledge and information with truth and understanding. Gemini explores the world, gathering information, making connections, and focusing primarily on the fundamental nature and expressions of duality. Sagittarius, on the other hand, seeks the unifying thread, the single idea, the great truth that connects all of creation with the creator.

The function of Gemini is to explore the environment and to make connections between different elements. Gemini relates to all forms of language and communication because words are simply ways of drawing a connection between ideas and objects. Gemini is fascinated by everything and has an absolutely insatiable curiosity about the world. Gemini is constantly gathering information and ideas, and exploring every possible facet and permutation of any situation. Inherent in the energy of Gemini is the concept of duality—Gemini will explore both extremes of any situation in an attempt to understand how the two opposite concepts relate to each other. Gemini, however, is so completely focused on the details that it lacks any kind of perspective. Gemini is always gathering information, but lacks the focus and attention span to make use of the raw data, to discover common themes, and to discover where the details fit in the bigger picture.

Sagittarius, on the other hand, is entirely concerned with the big picture. Sagittarius energy is focused and one-pointed, and always dedicated to discovering the ultimate truth. While Gemini explores the lower mind, Sagittarius operates on the level of the higher mind, in the realm of theory and philosophy, spirituality and theology. Where Gemini seeks to explore duality, Sagittarius wants to resolve it, integrating the opposite sides into a unified whole. Sagittarius is symbolized by the centaur, which merges our dual animal and human nature. Even though Sagittarius has focus where Gemini does not, Sagittarius too can lose perspective and become so obsessed with discovering the truth that it can adopt the idea that the end will always justify the means. While pursuing an understanding of the laws of the universe, Sagittarius can often forget the laws of man, inadvertently hurting the feelings of other individuals who may have different perspectives on their own personal truths.

The Gemini/Sagittarius nodal axis teaches us to find the balance between the lower and higher minds. We must always maintain our curiosity and flexibility, but it must also be guided by a higher understanding and philosophy. Searching for an understanding of the universal truths is important, but we must also discover how to apply those truths on a smaller scale in our daily lives—we must be able to communicate these truths to ourselves and to others. Knowledge must always be tempered with understanding and perspective.

North Node Gemini/South Node Sagittarius

The gifts of the South Node in Sagittarius include a core understanding of our individual relationship to the universe and to society. The South Node in Sagittarius gives a very strong belief system, a fundamental and unifying philosophy of life that can be a great help in expressing and reinforcing our individual identity. The trap of the South

Node in Sagittarius, however, involves adhering too rigidly to this belief system and not being receptive to other people's ideas and beliefs. The trap of the South Node in Sagittarius can result in a "holier than thou" attitude, as well as a tendency to talk the talk, but not walk the walk. While the South Node in Sagittarius most certainly has important information to share with the world, the only way to communicate these truths effectively is by working with the North Node in Gemini.

The North Node in Gemini can teach us how to apply the Sagittarius truths and beliefs to our daily life and environment. And perhaps, more importantly, through exploring the issues of duality that are always a part of Gemini energy, the South Node in Sagittarius can discover that the universal truth can appear in many different forms and express in many different and often contradictory ways.

North Node in Gemini in the 1st House/
South Node in Sagittarius in the 7th House

With your South Node in Sagittarius in the 7th house, you will encounter your South Node gifts of a dedication to truth, to personal freedom, and to exploring the more universal principles in the world through your relationships. One of the challenges of the 7th house, however, is that we tend to give away planets in this house, projecting them on others. You may feel that you lack the Sagittarius qualities, and therefore you may tend to attract people into your life who embody them. Until you are able to accept the gifts of the South Node in Sagittarius as being a part of you, you will experience and encounter them through your relationships. You may tend to attract and be attracted to people with a strong sense of purpose and clearly defined beliefs. How well you enjoy relating to these individuals will depend on to what degree you have owned, accepted, and integrated the Sagittarius South Node gifts into your life. In relationships, our partners reflect back to us things about ourselves that we are not yet ready to recognize. So if you find that you are frustrated by the number of didactic, tactless hypocrites in your life, it's likely that you have fallen into the trap of the South Node in Sagittarius in the 7th house and these relationships are your wake-up call to start integrating your beliefs and your philosophies into your daily life.

Working with your North Node in Gemini in the 1st house can help you balance this energy. However, we face a similar challenge with planets in the 1st house—while we tend to project the 7th house on others, we tend to overlook planets in the 1st house because they are such a fundamental part of who we are as an individual. The qualities of the North Node in Gemini—the curiosity, the quick intelligence, and above all the ability to take the larger Sagittarius concepts and translate them into everyday situations—are at your disposal; you simply need to cultivate a more conscious awareness of them. Your relationships will always be a source of information to you, showing you the ideas, concepts, and beliefs that you must now incorporate into your life and share with others.

North Node in Gemini in the 2nd House/
South Node in Sagittarius in the 8th House

With your South Node in Sagittarius in the 8th house, you will encounter and experience your South Node gifts of a dedication to truth, to personal freedom, and to exploring the more universal principles in the world when you are experiencing close emotional connections with other individuals. The 8th house is where we go to find a sense of emotional and soul security and self-worth. We seek a sense of inner peace through letting go of our physical and material attachments and merging with another individual on a deep, healing, and transformational level. The 8th house is also related to shared resources, and, like the 7th house, we often project our 8th-house planets on others, experiencing them through relationships. You much prefer to share your beliefs, philosophies, and experiences with another individual. You long for a partner to accompany you on your spiritual quest for truth, and so long as your partner's truth is compatible with your own, the shared experience can be powerful and energizing for both of you. The trap of the South Node in Sagittarius in the 8th house, however, is to become overly concerned with finding someone with whom to share your journey. If you are unable to find like-minded partners, you may begin to question your own truth.

The key to balancing this energy is to work with the North Node in Gemini in the 2nd house. The 2nd house represents our personal resources; anything that we can call our own belongs in the 2nd house, and everything in the 2nd house helps reinforce our sense of individuality. Our possessions, resources, skills, and talents, as well as our physical body and senses, help define and support our identity as an individual. With the North Node in Gemini in the 2nd house, you are encouraged to recognize the North Node lessons of duality, flexibility, and curiosity as valuable resources. Remember that the ultimate purpose of the North Node in Gemini is to break down the big Sagittarius concepts and discover how they apply to the smaller details in life; and to communicate these concepts. The North Node in Gemini in the 2nd house teaches that you do not have to rely on the support or approval of others in order to apply and communicate your truth. When you have that support, it's wonderful, but it's not a requirement.

North Node in Gemini in the 3rd House/
South Node in Sagittarius in the 9th House

With your South Node in Sagittarius in the 9th house, you encounter your gifts of a dedication to truth, to personal freedom, and to exploring the more universal principles in the world on a more philosophical and intellectual level. The 9th house is where we expand our understanding of the world and how we relate to it as an individual, and it includes philosophy, higher education, religion, and long journeys. Generally, the 9th house relates to more structured approaches to experience—particularly a more structured and ordered approach to spirituality as the 9th house relates to organized religion and to all of the attendant rules, strictures, pomp, and ritual to be found there. The trap of the South Node in Sagittarius in the 9th house is to become so focused on dogma that you lose sight of the true value of what it is that you seek to understand. You may tend to become so focused on principle, structure, and theory that you find it difficult to make use of anything you have learned in a practical sense. This energy can often manifest as religion without spirituality.

The key to balancing this energy is to work with the North Node in Gemini in the 3rd house. The 3rd house relates to our familiar environment and to how we communicate, reason, and make connections within that environment. The 3rd house also relates to our individual spiritual practices and beliefs—the ways that we connect to our spirituality outside of the structures of organized religion. Working with your North Node in Gemini in your 3rd house can help you find practical applications for your experience of truth. More than that, you can also use your North Node to discover how to connect with your individual spirituality, and experience your connection with the universe on a daily basis and without needing the trappings and dogma of organized religion as a guide.

North Node in Gemini in the 4th House/
South Node in Sagittarius in the 10th House

With your South Node in Sagittarius in the 10th house, you encounter your South Node gifts of a dedication to truth, to personal freedom, and to exploring the more universal principles in the world as a part of your public self. The 10th house is where we seek to create a tangible manifestation of our individual identity. It is where we want to make our mark on the world, and where we make our contributions to society as an individual. The 10th house is often associated with our career, but more accurately it is our life path, which is often quite different from our job. Your focus, honesty, and quest for truth will all serve you well in your public life. The trap of the South Node in Sagittarius in the 10th house, however, is to become entirely focused on the big picture, to devote so much of your energy and attention to your personal quest for truth that you lose sight of the more important details in your life, such as your family. Your drive towards personal freedom can result in your complete isolation.

The key to balancing this energy is to work with the North Node in Gemini in the 4th house. While the 10th house is the most public part of the chart, and also the most focused on individuality, the 4th house is the most private part of the chart, and the area where we are the most connected to others. The 4th house is our home; it is where we connect with our families (both of origin and of choice); but more than this, the 4th house represents our roots, our past, our ancestry, and our tribal heritage. It is the foundation on which we build our individuality and our public self. The North Node in Gemini in the 4th house indicates that communicating and sharing ideas and information with your family is extremely important to your spiritual path. The lessons and truths that you encounter in your public life must be shared with your family; they must be broken down into smaller, more understandable and practical pieces, and they must become an integrated part of your core emotional and spiritual foundation.

North Node in Gemini in the 5th House/
South Node in Sagittarius in the 11th House

With your South Node in Sagittarius in the 11th house, you encounter your South Node gifts of a dedication to truth, to personal freedom, and to exploring the more universal principles in the world when you socialize with your friends and peers. The 11th house is where we seek a sense of social and intellectual security. In the 11th house we want to belong to a social community. Within this community, we learn appropriate behavior standards, we participate in group creativity, and we learn how to receive and accept the love of others. Your friends are likely to share common beliefs and ideals with you, and this support, these common causes, often form the core foundation of your friendships. The trap of the South Node in Sagittarius in the 11th house is to assume that shared beliefs are the only basis for your friendships, and to choose to abandon your friends when they hold beliefs that are different from yours.

The key to balancing this energy is to work with the North Node in Gemini in the 5th house. The 5th house is where we search for security in our identity as an individual. Everything that we can do that makes us feel special and unique, that makes us feel like we deserve the attention and acknowledgment of others, is found in the 5th house—that is why the 5th house is associated with such a mixed bag of concepts including children, the arts, gambling, and love affairs. The combination of the Gemini energy of the North Node and the exceptionally fun nature of the 5th house means that the first lesson for you to learn is to lighten up and become more playful and flexible. The trap of the South Node in Sagittarius can make you so rigid in your beliefs that you are unable to tolerate any differences of opinion. While experiencing the freedom and playfulness of the North Node in Gemini in the 5th house, you are apt to discover that even the ultimate "truth" has many diverse facets when actually applied to life.

North Node in Gemini in the 6th House/
South Node in Sagittarius in the 12th House

With your South Node in Sagittarius in the 12th house, you may not be consciously aware of your gifts of a dedication to truth, to personal freedom, and to exploring the more universal principles in the world, even though they are very evident to other people. The 12th house is like our shadow: we can't see it directly because we're usually facing the light (of consciousness, that is), but it's very visible to everyone else. Our awareness of our 12th house seems to come from our unconscious and our subconscious. It probably doesn't occur to you how strongly you embody the quest for truth and honesty in all things, much less that these qualities are not always present in everyone else. The trap of the South Node in Sagittarius in the 12th house is that your unconscious drive for the truth can deteriorate into a need to always be right. You may be dismissive towards those who take a different view of the world. If you are not able to persuade others to convert to your beliefs, you will simply stop associating with them, which can result in extreme isolation.

The key to balancing this energy is to work with the North Node in Gemini in the 6th house. While the 12th house relates to our emotional and soul needs, the 6th house relates to our routine physical needs. The 6th house is related to illness because if we're not taking care of our body with proper nutrition, rest, and exercise on a daily basis, we will get sick. The 6th house is also related to our job and work environment, which is an essential part of maintaining our physical existence. With the North Node in Gemini in the 6th house, you must learn to take your understanding of the world, your approach to life, and your dedication to uncovering the ultimate truth, and apply them to your daily life. It's not enough to simply have a meditative spiritual life; you must find ways to apply these philosophies and teachings in your more mundane routines. These frequently include your job and your workplace, as the 6th house tends to be filled with co-workers much of the time. You will encounter many different beliefs and ideas through interacting with your co-workers. The North Node in Gemini in the 6th house encourages you to discover how your understanding of universal truth is still present even in these seemingly unconnected and unrelated concepts.

North Node in Gemini in the 7th House/
South Node in Sagittarius in the 1st House

With your South Node in Sagittarius in the 1st house, your gifts of a dedication to truth, to personal freedom, and to exploring the more universal principles in the world are very much a part of your sense of self; you access them naturally and may not even be consciously aware of them. The 1st house contains everything that we identify as being a fundamental part of who we are. In fact, it's quite difficult to gain perspective on the contents of the 1st house because it's too close to us. You may tend to define yourself in terms of your beliefs and your need for personal freedom. The trap of the South Node in Sagittarius in the 1st house is that just as you may not be able to gain perspective on the gifts of Sagittarius, you may also find it challenging to gain perspective on the less skillful expressions of Sagittarius energy. You may become so focused on your personal quest for truth that you wind up isolated from the rest of society; and of course if you focus exclusively on the big picture, you'll miss the important moments in life.

The key to balancing this energy is to work with the North Node in Gemini in the 7th house. The 7th house is the house of one-to-one relationships, and the biggest challenge of the 7th house is that we tend to give away the planets in this house, projecting them on others and experiencing the energies and lessons of these planets through our relationships. Working with the North Node in Gemini may be a challenge, because you may feel that you lack the qualities of Gemini, and as a result you will tend to attract people to you who embody these qualities. These relationships will help you learn how to accept and integrate the lessons of the North Node in Gemini in the 7th house. You may find that you tend to attract and be attracted to people who embody the curious, quick, intelligent, charming, and, above all, dual energy of your Gemini North Node. Rather than falling into the trap of your Sagittarius South Node in the 1st house and dismissing these people as being too flighty, these relationships are meant to teach you how to pay more attention to the details in your life. When you allow yourself to play with these partners, to share in their simple joy and curiosity, you can begin to recognize that the truth can be found just as easily by paying attention to the little things in life as it can by pursuing a cosmic identity quest.

North Node in Gemini in the 8th House/
South Node in Sagittarius in the 2nd House

With your South Node in Sagittarius in the 2nd house, your gifts of a dedication to truth, to personal freedom, and to exploring the more universal principles in the world are important skills and resources to you. The 2nd house represents our personal resources; anything that we can call our own belongs in the 2nd house, and everything in the 2nd house helps reinforce our sense of individuality. Our possessions, resources, skills, and talents, as well as our physical body and senses, help define and support our identity as an individual. The gifts of the South Node in Sagittarius in the 2nd house help support and reinforce your sense of self and your understanding of your place in the universe, and these beliefs, concepts, and personal truths provide you with a great deal of security. The trap of the South Node in Sagittarius in the 2nd house, however, is the tendency to become overly confident of the validity and value of your beliefs. When Sagittarius becomes too dogmatic and rigid, you will often experience reversals in your fortunes, because your financial resources rely on your ability to stay flexible.

The key to balancing this energy is to work with the North Node in Gemini in the 8th house. The 8th house is where we go to find a sense of emotional and soul security and self-worth. We seek a sense of inner peace through letting go of our physical and material attachments and merging with another individual on a deep, healing, and transformational level. The 8th house is also related to shared resources, and, like the 7th house, we often project our 8th-house planets on others, experiencing them through relationships. The North Node in Gemini in the 8th house encourages you to share your ideas and beliefs with a partner, and to be open to the fact that your partner may have a different perspective and experience of the "truth." The 8th-house experience of merging with a partner will afford you an entirely different perspective, and a new appreciation for your beliefs, as well as for the beliefs of others.

North Node in Gemini in the 9th House/
South Node in Sagittarius in the 3rd House

With your South Node in Sagittarius in the 3rd house, your gifts of a dedication to truth, to personal freedom, and to exploring the more universal principles in the world will tend to manifest in your ability to communicate and experience your immediate environment. The 3rd house represents the world that is most familiar to us, and the ways that we make sense of the world through logic, reason, and language. The 3rd house also relates to our individual spiritual practices and beliefs—the ways that we connect to our spirituality outside of the structures of organized religion. You are likely to have formed many of your core beliefs and perspectives at a relatively young age. While your philosophies, beliefs, and ideals and your dedication to personal freedom may serve you well when contained within the bounds of your home environment, ultimately this will not be enough for you. Even if you succumb to the trap of the South Node in Sagittarius in the 3rd house by choosing to be a "big fish in a small pond" and limiting your life to the familiar situations where your beliefs and perceptions will not be challenged, ultimately you will find this unfulfilling.

The key to balancing this energy is to work with the North Node in Gemini in the 9th house. The 9th house is where we expand our understanding of the world and how we relate to it as an individual, and it includes philosophy, higher education, religion, and long journeys. While the 9th house certainly shares many concepts with Sagittarius, the energy that you will encounter in the 9th house is the North Node in Gemini, and this encourages you to take a varied and unusual approach to your explorations. The North Node in Gemini in the 9th house isn't interested in pursuing the one great truth; instead, it's driven to sample and examine the widest range of new concepts and ideas, making connections between them and discovering how the frequently contradictory approaches to understanding the universe still share common themes and core concepts. The process of exploring these new intellectual and spiritual pursuits will in turn expand your 3rd-house world as you gain a new perspective on your old beliefs.

North Node in Gemini in the 10th House/
South Node in Sagittarius in the 4th House

With your South Node in Sagittarius in the 4th house, your gifts of a dedication to truth, to personal freedom, and to exploring the more universal principles in the world are closely related to your private life and family connections. The 4th house represents our home, our families (of choice and of origin), and our connection to our past through our ancestors and tribal heritage. The 4th house is our foundation—the rock on which we build our individual life. It supports us, and through the foundation of the 4th house, we are able to achieve the public, individual accomplishments of the 10th house. Your South

Node in Sagittarius gifts are a fundamental part of your soul identity. The universal truths that you remember on both an unconscious and a conscious level guide you and reveal to you your life's path. The trap of the South Node in Sagittarius in the 4th house, however, is that you may expect that the way your beliefs are shared and play out in your personal life should translate to the rest of the world. When you realize that setting out on your own means encountering and experiencing other people's truths, you may decide to stay isolated, connected to your family, and in an environment where you can be certain that your beliefs and understanding of truth will not be called into question by others.

The key to balancing this energy is to work with the North Node in Gemini in the 10th house. The 10th house is the most public part of the chart. It relates to our career and our life path; it is where we as individuals find our most significant accomplishments. The North Node in Gemini in the 10th house indicates that your life path includes the ability to communicate your South Node in Sagittarius truths to others. But more than that, the North Node in Gemini teaches that you must recognize that the truths you hold dear do indeed run as common threads through everything in your life; that even when you work with different and often opposing points of view, that each side of the argument, each facet of the situation, is nonetheless guided and shaped by universal law. With the North Node in Gemini in the 10th house, your discovery of the infinite and unexpected ways in which universal truth manifests is closely tied in with your life path. Finding a balance between your North and South Nodes will align your life path with your spiritual path.

North Node in Gemini in the 11th House/
South Node in Sagittarius in the 5th House

With your South Node in Sagittarius in the 5th house, you encounter your gifts of a dedication to truth, to personal freedom, and to exploring the more universal principles in the world whenever you explore and express your personal creativity. The 5th house is where we search for security in our identity as an individual. Everything that we can do that makes us feel special and unique, that makes us feel like we deserve the attention and acknowledgment of others, is found in the 5th house—that is why the 5th house is associated with such a mixed bag of concepts including children, the arts, gambling, and love affairs. Freedom of expression is extremely important to you. Your creative process is the path you choose to explore your spirituality and your relationship to the universe, and the truths you discover along the way will tend to be of a very personal nature. The trap of the South Node in Sagittarius in the 5th house is the tendency to become too isolated from the rest of the world; to make your creative and spiritual quest one that is so completely personal that you never feel the need or the desire to share your creativity with others, either because you don't believe that you need the support of anyone else, or because you are afraid that others will not accept your personal truth as their own truth.

The key to balancing this energy is to work with the North Node in Gemini in the 11th house. The 11th house is where we seek a sense of social and intellectual security. In the 11th house we want to belong to a social community. Within this community, we learn appropriate behavior standards, we participate in group creativity, and we learn how to receive and accept the love of others. With your North Node in Gemini in the 11th house, you are actively encouraged to share your ideas, beliefs, and self-expression with your friends, and to explore and sample the perspectives of others both independently and in relationship to your own points of view. Part of the goal here is to expand your horizons, to discover all of the myriad ways that your personal philosophy can be incorporated into different projects. By staying in touch with your South Node gifts, you won't ever lose sight of your own personal contribution to the group effort; and by incorporating the North Node in Gemini point of view, you will see how the common thread of your truth connects with the many different facets of the world.

North Node in Gemini in the 12th House/
South Node in Sagittarius in the 6th House

With your South Node in Sagittarius in the 6th house, you encounter your gifts of a dedication to truth, to personal freedom, and to exploring the more universal principles in the world in your daily routines. The 6th house contains everything that we must do on a daily basis to maintain our physical existence. The 6th house is related to illness because if we're not taking care of our body with proper nutrition, rest, and exercise on a daily basis, we will get sick. The 6th house is also related to our job and work environment, which is an essential part of maintaining our physical existence. Your philosophy of life and your passion for the truth are very much a part of your everyday existence. You are always open to new ideas and experiences, although you may tend to be a bit more open to ideas and experiences that fit easily within your core belief system. The challenge and the trap of the South Node in Sagittarius in the 6th house is that your South Node gifts can become such a part of your routine that you take them for granted, and they may cease to be special to you. You may tend to follow the most literal and narrow interpretations of your beliefs, and lose sight of the true spirit of them. From there, it's a short fall into the "holier than thou" aspects of the South Node in Sagittarius.

The key to balancing this energy is to work with the North Node in Gemini in the 12th house. Where the 6th house relates to how we take care of our physical health, the 12th house is how we take care of our spirit and our soul. In the 12th house we encounter our need to feel a part of a spiritual and emotional community; to let go of our individuality and to merge once again with the universe. We retreat to the 12th house whenever we need to take a break from the demands of our daily routines. With the North Node in Gemini in your 12th house, you must learn to rediscover your natural curiosity and childlike innocence with respect to spiritual matters. This energy tends to be more abstract than practical. In order to work with your North Node, you will need to take a fresh look at your beliefs and the things that you consider to be true, and to allow yourself to question them as well as to explore ideas and beliefs that are contrary to the ones you currently hold. Gemini's essence is duality, and by working with your North Node in Gemini in the 12th house, you can discover how your South Node in Sagittarius appreciation for the truth can be a common element in seemingly disparate and even opposing ideas.

North Node Sagittarius/South Node Gemini

The gifts of the South Node in Gemini include an insatiable curiosity about the world, a quick mind that is open to all forms of new ideas, and a playful, youthful outlook. The South Node in Gemini loves variety and is always searching for new ideas and information. The trap of the South Node in Gemini, however, is a lack of focus and difficulty integrating the vast stores of information accumulated. The South Node in Gemini can be very much a "Jack of all trades and a master of none." The key to balancing this energy is to work with the lessons of the North Node in Sagittarius and to look for the common thread that links the dual concepts. Gemini already looks for connections, and the Sagittarius influence simply helps Gemini look for connections on a broader and much larger scale. Gemini collects information, but working with the focus of Sagittarius, that information can be catalogued and synthesized. Working with the Sagittarius North Node takes the Gemini information and distills from it an understanding about how the universe works, and our role in the greater scheme of things. Higher education, philosophy, travel, and cross-cultural studies are some of the paths that the Sagittarius North Node may take in the search for truth and a more complete understanding of cosmic law. A Gemini South Node is a strong indication that you have a message to share with the world; by balancing the energy of Gemini with the focus and perspective of Sagittarius, that message will become much clearer to you.

North Node in Sagittarius in the 1st House/
South Node in Gemini in the 7th House

With your South Node in Gemini in the 7th house, you will encounter your South Node gifts of communication, rapid assimilation of information, and a fundamental curiosity about things through your relationships. One of the challenges of the 7th house, however, is that we tend to give away planets in this house, projecting them on others. You may feel that you lack the Gemini qualities, and therefore you may tend to attract people into your life who embody them. Until you are able to accept the gifts of the South Node in Gemini as being a part of you, you will experience and encounter them through your relationships. You may feel that you lack the ability to be flexible, to be curious and playful, and therefore you both admire this quality in others and tend to attract people into your life who embody and express these qualities. Owing to the dual nature of Gemini, you may need to look at your relationships as a collective rather than on an individual basis to see the true scope of the variety of different individuals who contribute to your life. The trap of the South Node in Gemini in the 7th house is to move through these relationships, always keeping them on the most intellectual and social level, and never looking at what bigger lessons your relationships are trying to teach you. All relationships are mirrors, reflecting aspects of ourselves back to us so that we can ultimately accept and own these parts of ourselves.

Working with your North Node in Sagittarius in the 1st house can help you balance this energy. However, we face a similar challenge with planets in the 1st house—while we tend to project the 7th house on others, we tend to overlook planets in the 1st house because they are such a fundamental part of who we are as an individual. With the North Node in Sagittarius in your 1st house, your first challenge is to recognize that your ability to see the big picture, to understand and assimilate theoretical and philosophical concepts and how they apply to the little things in life, is not an ability shared by everyone. True, it is very much a part of who you are, and your perceptions and experience of the world are very much influenced by this perspective, but you must still become aware of it on a conscious level in order to make use of it. Perspective is what you need most, as the truth that you are seeking, the common thread that runs through all of your various and varied relationships, can only be seen by looking at your relationships as a whole. Each one of your relationships reflects back a tiny part of yourself that you must learn to integrate by considering who you are when you relate to each individual: How do they perceive you? Your North Node in Sagittarius in the 1st house gives you the ability to connect the Gemini dots and see the big picture, and in the process, get closer to the ultimate truth of your own unique identity.

North Node in Sagittarius in the 2nd House/
South Node in Gemini in the 8th House

With your South Node in Gemini in the 8th house, you will encounter and experience your South Node gifts of communication, rapid assimilation of information, and a fundamental curiosity about things when you are experiencing close emotional connections with other individuals. The 8th house is where we go to find a sense of emotional and soul security and self-worth. We seek a sense of inner peace through letting go of our physical and material attachments and merging with another individual on a deep, healing, and transformational level. The 8th house is also related to shared resources, and, like the 7th house, we often project our 8th-house planets on others, experiencing them through relationships. Gemini is not very comfortable with extended emotional connections, and the South Node in Gemini in the 8th house will tend to make up for the short duration of these connections by experiencing these connections with a large number of individuals. One way that this energy can manifest is by sharing certain personal secrets with others—forming bonds based on quick bursts of intimacy and shared pain or trauma. The trap of the South Node in Gemini in the 8th house is to become addicted to these connections; to use our wounds and our personal trauma as currency, exchanging it with others for a sense of self-worth and the feeling that our pain and emotional trauma is justified.

The key to balancing this energy is to work with the North Node in Sagittarius in the 2nd house. The 2nd house represents our personal resources; anything that we can call our own belongs in the 2nd house, and everything in the 2nd house helps reinforce our sense of individuality. Our possessions, resources, skills, and talents, as well as our physical body and senses, help define and support our identity as an individual. One of the truths of the North Node in Sagittarius in the 2nd house is that you can become responsible for your own sense of self-worth. You must learn how to transcend the details, to look at the different types of 8th-house connections and relationships that you have, and to discover what the common theme of these relationships is. The answer to that question is also the answer to the question of what is it that you value most in life? You must move from your experiences with shared resources and shared intellectual intimacy and move towards the truth of the North Node in Sagittarius in the 2nd house: What are your true skills and resources? What truly belongs to you?

North Node in Sagittarius in the 3rd House/
South Node in Gemini in the 9th House

With your South Node in Gemini in the 9th house, you encounter your gifts of communication, rapid assimilation of information, and a fundamental curiosity about things on a more philosophical and intellectual level. The 9th house is where we expand our understanding of the world and how we relate to it as an individual, and it includes philosophy, higher education, religion, and long journeys. Your South Node gifts indicate that you have a tremendous capacity for exploring your 9th house, and that you have a passion for new information and new ideas. The trap of the South Node in Gemini in the 9th house, however, is that you will tend to learn a little bit about a lot of different things, and never care to look beneath the surface or to explore the deeper aspects of the ideas you discover. You may tend to be in constant motion, jumping from one idea to the next, prompted by whim and free association, and find it difficult to pursue one line of thought or one idea to its conclusion.

The key to balancing this energy is to work with the North Node in Sagittarius in the 3rd house. The 3rd house relates to our familiar environment, and to how we communicate, reason, and make connections within that environment. The 3rd house also relates to our individual spiritual practices and beliefs—the ways that we connect to our spirituality outside of the structures of organized religion. With the North Node in Sagittarius in the 3rd house, you must take the ideas and philosophies that you find in the 9th house and begin to apply them to your everyday life. You must begin to look for the ways that these different ideas and beliefs manifest in your daily environment, and discover how they shape your perceptions of the world. As you do this, you will begin to understand the different ideas on a deeper level, and the unifying connections between the different and often opposing points of view will become more evident to you. As you focus more on these connections, you will discover that you are able to find your personal truth without ever having to leave your familiar environment. It's not necessary to travel great distances to find wisdom and spirituality; we simply need to know how to look for it where we are.

North Node in Sagittarius in the 4th House/
South Node in Gemini in the 10th House

With your South Node in Gemini in the 10th house, you encounter your South Node gifts of communication, rapid assimilation of information, and a fundamental curiosity about things as a part of your public self. The 10th house is where we seek to create a tangible manifestation of our individual identity. It is where we want to make our mark on the world, and where we make our contributions to society as an individual. The 10th house is often associated with our career, but more accurately it is our life path, which is often quite different from our job. You may have a tendency to scatter your energy and to be involved in many different types of public activities, either personally or through your career. You may have a great many different professional interests, and choose to divide your time and energy so that you can work with all of them rather than choosing a single path. The trap of the South Node in Gemini in the 10th house is that you can easily find that you've bitten off more than you can chew. The endless fascination and curiosity of Gemini can keep you in constant motion, always learning new things and new ways of understanding the world. This scattering of energies, however, can begin to take its toll, especially on a personal level.

The key to balancing this energy is to work with the North Node in Sagittarius in the 4th house. While the 10th house is the most public part of the chart, and also the most focused on individuality, the 4th house is the most private part of the chart, and the area where we are the most connected to others. The 4th house is our home; it is where we connect with our families (both of origin and of choice); but more than this, the 4th house represents our roots, our past, our ancestry, and our tribal heritage. It is the foundation on which we build our individuality and our public self. With the North Node in Sagittarius in the 4th house, the North Node process of seeking truth and personal freedom is a very private one. The truth and understanding that you seek is intended to help you reconnect with your past, your ancestors, and with the ultimate source of all life. For you, the quest for truth is something that you must seek within yourself rather than outside of yourself. It is important that you learn how to take time for yourself, to work with your North Node in the 4th house, and integrate the vast amount of information that you can obtain through your South Node, discovering the common thread that connects and binds all of the opposites. By following this thread, you can connect with your own spirituality and discover your relationship to the rest of creation.

North Node in Sagittarius in the 5th House/
South Node in Gemini in the 11th House

With your South Node in Gemini in the 11th house, you encounter your South Node gifts of communication, rapid assimilation of information, and a fundamental curiosity about things when you socialize with your friends and peers. The 11th house is where we seek a sense of social and intellectual security. In the 11th house we want to belong to a social community. Within this community, we learn appropriate behavior standards, we participate in group creativity, and we learn how to receive and accept the love of others. You may find that you easily attract a wide variety of friends and enjoy spending time in social situations. Other individuals, and particularly groups of individuals, may hold an endless fascination for you, and you may find that you tend to have a full social calendar. The trap of the South Node in Gemini in the 11th house, however, is to stay focused on social interaction with friends and neglect developing your own sense of personal creativity.

The key to balancing this energy is to work with the North Node in Sagittarius in the 5th house. The 5th house is where we search for security in our identity as an individual. Everything that we can do that makes us feel special and unique, that makes us feel like we deserve the attention and acknowledgment of others, is found in the 5th house—that is why the 5th house is associated with such a mixed bag of concepts including children, the arts, gambling, and love affairs. The North Node in Sagittarius in the 5th house can be experienced through the arts, and particularly through exploring and experimenting with many different art forms from many different cultures. Sagittarius seeks to expand and to explore in order to find the universal truth, and in the 5th house, the universal truth that you seek is the truth of your own creative gifts. The key to the 5th house is active focus and participation. The South Node in Gemini in the 11th is perfectly happy to appreciate other people's creative achievements; the North Node in Sagittarius in the 5th needs to, if not be the creator, at least be able to understand on a personal level the creative process.

North Node in Sagittarius in the 6th House/
South Node in Gemini in the 12th House

With your South Node in Gemini in the 12th house, you may not be consciously aware of your gifts of communication, rapid assimilation of information, and a fundamental curiosity about things, even though they are very evident to other people. The 12th house is like our shadow:

we can't see it directly because we're usually facing the light (of consciousness, that is), but it's very visible to everyone else. Our awareness of our 12th house seems to come from our unconscious and our subconscious. On the positive side, the Gemini energy in your 12th house makes you very receptive to new ideas and concepts; you display a natural curiosity and thirst for new information. The trap of the South Node in Gemini in the 12th house, however, involves being too open, and collecting so much new information that you are simply unable to process it all. Of course, since you are probably not aware of what has happened, you may experience this as feeling overwhelmed and scattered for no apparent reason.

The key to balancing this energy is to work with the North Node in Sagittarius in the 6th house. While the 12th house relates to our emotional and soul needs, the 6th house relates to our routine physical needs. The 6th house is related to illness because if we're not taking care of our body with proper nutrition, rest, and exercise on a daily basis, we will get sick. The 6th house is also related to our job and work environment, which is an essential part of maintaining our physical existence. The North Node in Sagittarius in the 6th house encourages you to focus on the big picture as you make your way through your daily routines. Part of the process involves integrating the spiritual information that you gather through your South Node in Gemini in the 12th house and discovering how it all fits into the greater unifying truths of your life. Since one of the most fundamental things that we must do to maintain our physical existence is to make a living, much of the energy of the 6th house tends to play out in our jobs and through our relationships with our co-workers. You must learn how to discover and express your personal truth in all things, and to be able to express and defend that truth when necessary. Maintaining a connection with your South Node in Gemini gifts will allow you to keep an open mind when others have different beliefs, but your North Node in Sagittarius encourages you to stay focused on your own path and not to be distracted by the paths of others.

North Node in Sagittarius in the 7th House/
South Node in Gemini in the 1st House

With your South Node in Gemini in the 1st house, your gifts of communication, rapid assimilation of information, and a fundamental curiosity about things are very much a part of your sense of self; you access them naturally and may not even be consciously aware of them. The 1st house contains everything that we identify as being a fundamental part of who we are. In fact, it's quite difficult to gain perspective on the contents of the 1st house because it's too close to us. Your curiosity, your intelligence, and your social skills and charm all come so naturally to you that you may not even realize (or care) that these are qualities that are not shared by everyone. The trap of the South Node in Gemini in the 1st house, however, is to become a dilettante; to dabble in so many different paths, to spend too much energy polishing and expressing the different facets of your personality, that you are never able to fully integrate and discover the unifying truth of who you are.

The key to balancing this energy is to work with the North Node in Sagittarius in the 7th house. The 7th house is the house of one-to-one relationships, and the biggest challenge of this house is that we tend to give away the planets in this house, projecting them on others and experiencing the energies and lessons of these planets through our relationships. Working with the North Node in Sagittarius may be a challenge, because you may feel that you lack the qualities of Sagittarius, and as a result, you will tend to attract people to you who embody these qualities. These relationships will help you learn how to accept and integrate the lessons of the North Node in Sagittarius in the 7th house. You may find that you both attract and are attracted to people who have a very strong sense of self; who are secure in their beliefs and passionate about their quest for truth and self-realization. These relationships can often be challenging, as these Sagittarian individuals have little time and even less tolerance for frivolous games. They will want to discover the unifying truth of who you are, and will not settle for experiencing your different facets, no matter how well polished they may be. Ultimately, these relationships are reflecting back to you the lessons of your North Node in Sagittarius in the 7th house. It is through your relationships, through observing how you relate to your partners and how they seem to perceive and respond to you, that you can begin to discover the constant elements that are present regardless of what facet of yourself you present, and therein lies the truth of your true self.

North Node in Sagittarius in the 8th House/ South Node in Gemini in the 2nd House

With your South Node in Gemini in the 2nd house, your gifts of communication, rapid assimilation of information, and a fundamental curiosity about things are important skills and resources to you. The 2nd house represents our personal resources; anything that we can call our own belongs in the 2nd house, and everything in the 2nd house helps reinforce our sense of individuality. Our possessions, resources, skills, and talents, as well as our physical body and senses, help define and support our identity as an individual. Because of the fundamentally dual nature of Gemini, what you may tend to value the most in life is having choices and options—even more than that, however, your greatest wish may be to "have your cake and eat it too." One of the disadvantages of actually making a choice between two or more options is that at that point, the remaining options are no longer open to you, something that your South Node in Gemini can't bear. While your flexibility is certainly a gift, when taken to the extreme it is also the trap of the South Node in Gemini in the 2nd house: that you may value your options and your choices so much that you are unable to commit to any one choice for long enough to see any real value from the resource.

The key to balancing this energy is to work with the North Node in Sagittarius in the 8th house. The 8th house is where we go to find a sense of emotional and soul security and self-worth. We seek a sense of inner peace through letting go of our physical and material attachments and merging with another individual on a deep, healing, and transformational level. The 8th house is also related to shared resources, and, like the 7th house, we often project our 8th-house planets on others, experiencing them through relationships. The North Node in Sagittarius in the 8th house encourages you to find an underlying truth, a unifying principle to all of your South Node in Gemini duality, and suggests that you may be able to find this truth through intimate relationships with others. By exploring the concepts of shared resources, you can learn how to make a choice and stay focused on one option for your own personal resources, but still maintain contact with the options you didn't choose vicariously through your partners. Through the give and take of intimate relationships, you can learn to experience both sides of your duality together, and ultimately discover the common thread that unites the two polarities into a coherent whole.

North Node in Sagittarius in the 9th House/ South Node in Gemini in the 3rd House

With your South Node in Gemini in the 3rd house, your gifts of communication, rapid assimilation of information, and a fundamental curiosity about things will tend to manifest in your ability to communicate and experience your immediate environment. The 3rd house represents the world that is most familiar to us, and the ways that we make sense of the world through logic, reason, and language. The 3rd house also relates to our individual spiritual practices and beliefs—the ways that we connect to our spirituality outside of the structures of organized religion. Your curiosity is primarily limited to what is familiar to you. The 3rd house relates to our early education and childhood experiences, and shares many qualities with the sign of Gemini. You may be quite comfortable in the world you have created for yourself; you know how things work, you are able to be playful and social, and you can always find something to keep your mind occupied. The trap of the South Node in Gemini in the 3rd house is to become too attached to the carefree, playful atmosphere; to never want to grow up and mature.

The key to balancing this energy is to work with the North Node in Sagittarius in the 9th house. The 9th house is where we expand our understanding of the world and how we relate to it as an individual, and it includes philosophy, higher education, religion, and long journeys. The process of the North Node in Sagittarius in the 9th house is one of maturing, of discovering that our limited understanding of our environment is no longer enough, and of setting out on a quest to discover more about the universe and our place in it. The South Node in Gemini gifts can serve you well on this journey; you simply need to apply your curiosity to 9th-house pursuits, and learn to keep an eye on the bigger picture and not to be as distracted by the details.

North Node in Sagittarius in the 10th House/
South Node in Gemini in the 4th House

With your South Node in Gemini in the 4th house, your gifts of communication, rapid assimilation of information, and a fundamental curiosity about things are closely related to your private life and family connections. The 4th house represents our home, our families (of choice and of origin), and our connection to our past through our ancestors and tribal heritage. The 4th house is our foundation—the rock on which we build our individual life. It supports us, and through the foundation of the 4th house, we are able to achieve the public, individual accomplishments of the 10th house. The South Node in Gemini in the 4th house is often an indication of a very diverse and unusual family history. The dual nature of Gemini can mean that you draw from a number of different cultures, religions, and heritages. Your home life can tend to be very social and playful, with great variety. The trap of the South Node in Gemini in the 4th house is to want to stay rooted to the familiar experiences of your home life. You may choose to participate in your family environment without ever truly exploring what it means to you. You may never question how the different cultures, beliefs, and ancestors in your family came together to create you, and how you as an individual combine elements of all of these different perspectives.

The key to balancing this energy is to work with the North Node in Sagittarius in the 10th house. The 10th house is the most public part of the chart. It relates to our career and our life path; it is where we as individuals find our most significant accomplishments. With the North Node in Sagittarius in the 10th house, you are challenged to discover who you are as an individual, and as you uncover this personal truth, to follow that truth as the guiding force of your life and career path. The process will require you to work closely with your South Node in Gemini in the 4th house, to look at the different lives, families, and cultures that joined together to create you—this is where your journey will begin. As you explore your heritage, and explore how you experienced that heritage as a child and within your family dynamic, you will begin to discover how these ideas, concepts, and beliefs make up your core foundation. As you build on this foundation, and accept the truth of it, you will begin to solidify your sense of self, and to connect with your role and place in the universe.

North Node in Sagittarius in the 11th House/ South Node in Gemini in the 5th House

With your South Node in Gemini in the 5th house, you encounter your gifts of communication, rapid assimilation of information, and a fundamental curiosity about things whenever you explore and express your personal creativity. The 5th house is where we search for security in our identity as an individual. Everything that we can do that makes us feel special and unique, that makes us feel like we deserve the attention and acknowledgment of others, is found in the 5th house—that is why the 5th house is associated with such a mixed bag of concepts including children, the arts, gambling, and love affairs. You may find that you have a wide variety of artistic and creative interests. Although the 5th house is often associated with the performing arts, the Gemini influence here could indicate that you enjoy writing, perhaps several different types of writing. Expressing yourself creatively is very important to the South Node in Gemini in the 5th house. The trap here, however, is the tendency to isolate yourself and to keep your perspective focused on yourself. Your creativity will tend to run dry very quickly unless you expand your experience and your understanding through your North Node in Sagittarius in the 11th house.

The 11th house is where we seek a sense of social and intellectual security. In the 11th house we want to belong to a social community. Within this community, we learn appropriate behavior standards, we participate in group creativity, and we learn how to receive and accept the love of others. Sagittarius seeks to expand and to explore in order to find the universal truth. With the North Node in Sagittarius in the 11th house, you will tend to find this truth through group activities and through spending time with friends and peers. As Sagittarius energy is also about cross-cultural experiences, you may find that you develop a very wide and diverse circle of friends from many different backgrounds. As you allow yourself to become part of a group, as you involve yourself in group activities with your friends, you will begin to discover the truth that you seek, the common theme and understanding that will revitalize your creative abilities.

North Node in Sagittarius in the 12th House/ South Node in Gemini in the 6th House

With your South Node in Gemini in the 6th house, you encounter your gifts of communication, rapid assimilation of information, and a fundamental curiosity about things in your daily routines. The 6th house contains everything that we must do on a daily basis to maintain our physical existence. The 6th house is related to illness because if we're not taking care of our body with proper nutrition, rest, and exercise on a daily basis, we will get sick. The 6th house is also related to our job and work environment, which is an essential part of maintaining our physical existence. You may tend to enjoy a great deal of variety in your daily life, and in fact you may feel that the entire concept of a "daily routine" is far too limiting for your taste. The social connections that you make are very important to you, and you may find that you enjoy a great deal of socializing and a spirit of playful camaraderie with your co-workers. In addition to our daily routines and responsibilities, the 6th house is where we look to serve others, to assist in making the world a better place. With your South Node in Gemini in your 6th house, sharing information and providing alternate views is how you look to be of service to others. The trap of the South Node in Gemini in the 6th house, however, is that you may tend to scatter your energy and to lack focus. The South Node in Gemini in the 6th house can devote all of its energy to investigating the details and individual components of everyday life, making connections and exploring different facets, but ultimately be unable to find a unifying theme; without that common thread, the details can become muddled and confusing.

The key to balancing this energy is to work with the North Node in Sagittarius in the 12th house. Where the 6th house relates to how we take care of our physical health, the 12th house is how we take care of our spirit and our soul. In the 12th house we encounter our need to feel a part of a spiritual and emotional community; to let go of our individuality and to merge once again with the universe. We retreat to the 12th house whenever we need to take a break from the demands of our daily routines. With the North Node in Sagittarius in your 12th house, part of your lesson may be to learn how to place yourself in the proper perspective with the universe by retreating from the buzz of your daily life, letting go of your fascination with the details and allowing yourself to step back and look at the big picture. Although the time and energy we spend in the physical realm is important, it is only a small piece of the puzzle. Allowing yourself the time to explore your unconscious self and your spirituality by working with your North Node in Sagittarius in your 12th house gives you the chance to glimpse a portion of your personal truth. With this new perspective to guide you, you will be able to work with your South Node in Gemini and discover how this higher perspective truly does connect and pervade all of the details of your daily life. Then you truly have something to communicate and share with others.

6

THE CANCER/CAPRICORN
NODAL AXIS

The Cancer/Capricorn nodal axis is the axis of responsibility. The purpose of this axis is to learn how to find a balance between our responsibility to take care of ourselves, and our responsibility to take care of others in the form of our duties to our families and to society. Both Cancer and Capricorn are concerned with responsibility and with meeting our fundamental needs in life. Cancer is concerned with making sure that our emotional and soul needs are met—that we feel nurtured, safe, and emotionally protected. Capricorn is concerned with making sure that the structures and laws of society are formed and maintained so that everyone is protected and supported. Although the more obvious need for balance between Cancer and Capricorn has to do with our personal/family obligations versus our professional obligations, within each sign is also an inherent need to find a balance between being selfish and selfless. Both signs struggle with the lesson that before we can take care of others, we have an obligation to take care of ourselves—we must always come first.

Cancer is where we seek to feel a sense of emotional and spiritual connection, to once again experience the safety and comfort we felt before we separated from the universal consciousness. On the highest level, Cancer understands that because we are all connected and part of all that is, when we nurture and care for others, we are also nurturing ourselves; and when we are being nurtured, we are also nurturing others. Because there are truly no boundaries or separations between us, there is no difference between giving love and receiving it. Cancer, however, does not always operate on this level, and all too often we forget the truth of our connections to each other, and buy into the illusion of ego and separation. When this occurs, Cancer can become obsessed with having its personal emotional and security needs met. This can manifest as exceptionally needy and dependent behavior, or as the eternal caretaker, the person who is always giving and never allows anyone else to give back in return for fear that if he or she stops giving, others will not give back.

Capricorn, on the other hand, is concerned with the needs of society rather than the needs of the individual. Being an earth sign, Capricorn operates on the material and practical level and is motivated by structure and support, whether these are physical (in the form of shelters) or social (in the form of laws and guidelines for acceptable behavior). In its highest manifestation, Capricorn understands our connection to all of creation as well, and therefore understands that when we take responsibility for protecting the structure of society and meeting the physical needs of others, we are also meeting our own physical and material needs. But Capricorn can also become ego-involved and focused only on personal gain and ambition, pursuing power and influence for its own sake at the cost of personal and individual happiness.

The Cancer/Capricorn nodal axis teaches us about our responsibilities to ourselves and to others. We must at once learn to balance between our family obligations and our responsibility to be a contributing member of society. But in each case, we must also learn what it means to be truly selfish—a concept that currently has a very negative connotation. Being truly selfish simply means being responsible for making sure that our individual fundamental needs are being met before we devote any of our resources to helping others meet their needs. Until we can help ourselves, we cannot be expected to help others. Once we are able to help ourselves, however, then we do have an obligation to assist others not simply by meeting their needs for them, but by helping them discover how they can take personal responsibility for their lives, and begin to meet their needs on their own.

North Node Cancer/South Node Capricorn

The South Node in Capricorn offers the gift of a strong sense of self-reliance, personal responsibility, and practicality. Capricorn is the builder of the zodiac, and the South Node in Capricorn indicates advanced skills in manifestation of your desires and ideas. You have a great reserve of self-reliance and independence at your disposal. The South Node in Capricorn also gives you the ability to understand your role and responsibility in society, and to plan and execute long-range goals efficiently. The trap of the South Node in Capricorn, however, is the belief that relying on other people for help and forming emotional connections is a sign of weakness. The South Node in Capricorn may have little time for personal concerns or for emotional and spiritual issues; you may tend to distance yourself from this type of situation by taking on jobs, careers, and responsibilities to others that consume much of your time and energy, leaving little time leftover for a personal life.

This energy can be balanced by working with the lessons of the North Node in Cancer, which will encourage you to knock down some of your walls and open your heart to emotional connections. Oftentimes the most difficult part of this is for you to accept that you don't always have to be in control and responsible; that you can let down your guard and allow other people to take care of you. And ultimately, of course, in order to be truly responsible to ourselves as individuals, we must experience these emotional connections in our lives.

North Node in Cancer in the 1st House/
South Node in Capricorn in the 7th House

With your South Node in Capricorn in the 7th house, you will encounter your South Node gifts of ambition, responsibility, and self-reliance through your relationships. One of the challenges of the 7th house, however, is that we tend to give away planets in this house, projecting them on others. You may feel that you lack the Capricorn qualities, and therefore you may tend to attract people into your life who embody them. Until you are able to accept the gifts of the South Node in Capricorn as being a part of you, you will experience and encounter them through your relationships. You may feel that you lack the ability to be practical, grounded, and responsible, and therefore you may tend to attract and be attracted to people who embody these qualities. This can manifest as partners who are controlling and authoritarian, who seem emotionally distant, and who are rigid in their agenda. You must learn to recognize that these relationships are meant to teach you to own your South Node in Capricorn and to learn to maintain and establish your own interpersonal boundaries so that you won't have to be limited by those of other people. The trap of the South Node in Capricorn in the 7th house is to buy into the illusions of these relationships, and to decide that you must become entirely self-sufficient and not rely on the support of others for any reason; or to go to the opposite extreme and decide that you will always need someone to take care of you and as a result never learn to take responsibility for your actions and your relationships.

Working with your North Node in Cancer in the 1st house can help you balance this energy. However, we face a similar challenge with planets in the 1st house—while we tend to project the 7th house on others, we tend to overlook planets in the 1st house because they are such a fundamental part of who we are as an individual. With Cancer in your 1st house, you have a tremendous ability to nurture, and an ability and need to connect with others on an emotional level. You must first cultivate a greater awareness of this fact, and then begin to apply it to your relationships. Your partners may always be the ones who provide more of the boundaries and structures in your relationships, but you must learn to focus on your ability to share emotional bonds within those relationships. Much of the important give and take in relationships has to do with being able to support and nurture each other on an emotional level. Maintaining an awareness of your South Node in Capricorn gifts can ensure that you don't overdo the emotional aspects and become too needy or dependent on your partners. Instead, you can discover the appropriate balance between individual responsibility and shared emotional needs.

North Node in Cancer in the 2nd House/
South Node in Capricorn in the 8th House

With your South Node in Capricorn in the 8th house, you will encounter and experience your South Node gifts of ambition, responsibility, and self-reliance when you are experiencing close emotional connections with other individuals. The 8th house is where we go to find a sense of emotional and soul security and self-worth. We seek a sense of inner peace through letting go of our physical and material attachments and merging with another individual on a deep, healing, and transformational level. The 8th house is also related to shared resources, and, like the 7th house, we often project our 8th-house planets on others, experiencing them through relationships. With your South Node in Capricorn in the 8th house, you are apt to be extremely responsible and practical when it comes to shared resources. Capricorn energy in the 8th house often makes for excellent bankers and investment advisors because these individuals take other people's resources quite seriously and maintain strict standards and boundaries. The trap of the South Node in Capricorn in the 8th house is to become obsessed with managing and building, and working with shared resources and other people's money. This not only results in your giving away much of your own control and power in these relationships, but it also leads to placing far too great an emphasis on the material plane.

The key to balancing this energy is to work with the North Node in Cancer in the 2nd house. The 2nd house represents our personal resources; anything that we can call our own belongs in the 2nd house, and everything in the 2nd house helps reinforce our sense of individuality. Our possessions, resources, skills, and talents, as well as our physical body and senses, help define and support our identity as an individual. The North Node in Cancer in the 2nd house indicates that one of your most important skills and resources is your ability to be guided by your heart; that you can put emotions and feelings first and give them priority over practical, material considerations. Knowing that you can always rely on the support of others, thanks to your South Node in Capricorn in the 8th house, gives you the freedom to place more emphasis on meeting your own emotional needs. Part of the lesson of the North Node in Cancer in the 2nd house is to recognize that you must allow yourself to be selfish at times, and place your own emotional needs over your feeling of responsibility and duty to others.

North Node in Cancer in the 3rd House/
South Node in Capricorn in the 9th House

With your South Node in Capricorn in the 9th house, you encounter your gifts of ambition, responsibility, and self-reliance on a more philosophical and intellectual level. The 9th house is where we expand our understanding of the world and how we relate to it as an individual, and it includes philosophy, higher education, religion, and long journeys. The South Node in Capricorn's drive to build a tangible identity places great emphasis on education and knowledge in the 9th house. Your ambition may be focused on intellectual and scholastic achievements, and your sense of identity can be enhanced and supported by the degrees you have earned or the cultures you have experienced. The trap of the North Node in Capricorn in the 9th house is to begin to define yourself through these accomplishments. In your pursuit of more knowledge and greater recognition, you will find it more difficult to reach your true objective of discovering who you are as an individual and what your place is in the universe.

The key to balancing this energy is to work with the North Node in Cancer in the 3rd house. The 3rd house relates to our familiar environment, and to how we communicate, reason, and make connections within that environment. The 3rd house also relates to our individual spiritual practices and beliefs—the ways that we connect to our spirituality outside of the structures of organized religion. The North Node in Cancer in the 3rd house encourages you to explore the emotional connections found in your own backyard, rather than devote your entire focus to building an identity out of abstract philosophical concepts. Part of the lesson here is to take the concept of your self that you have built through your exploration of the South Node in Capricorn in the 9th house, and to bring that back home and explore your need for emotional connections. You can learn to take comfort from the familiar and to be nurtured and supported emotionally by the smaller truths in life.

North Node in Cancer in the 4th House/
South Node in Capricorn in the 10th House

With your South Node in Capricorn in the 10th house, you encounter your South Node gifts of ambition, responsibility, and self-reliance as a part of your public self. The 10th house is where we seek to create a tangible manifestation of our individual identity. It is where we want to make our mark on the world, and where we make our contributions to society as an individual. The 10th house is often associated with our career, but more accurately it is our life path, which is often quite different from our job. You may find it very easy to take on the challenges of your career. You may find that you naturally gravitate towards positions and situations where you are responsible for making decisions that will impact others. Because of your ability to focus on the bottom line and to take action based on what is most practical, you may tend to be singled out as a leader and encouraged to take on positions and roles of even greater responsibility. The trap of the South Node in Capricorn in the 10th house, however, is that your public responsibilities may begin to overwhelm and overshadow your personal needs and your connections and responsibilities to your family.

The key to balancing this energy is to work with the North Node in Cancer in the 4th house. While the 10th house is the most public part of the chart, and also the most focused on individuality, the 4th house is the most private part of the chart, and the area where we are the most connected to others. The 4th house is our home; it is where we connect with our families (both of origin and of choice); but more than this, the 4th house represents our roots, our past, our ancestry, and our tribal heritage. It is the foundation on which we build our individuality and our public self. The challenge and lesson here is to realize that your first and fundamental responsibility is to yourself; only after you have taken care of your own needs can you take on responsibility for the needs and security of others. What you must learn to accept, however, is that your needs include the need for emotional connections and intimacy, and also that you may have to rely on and accept help from others in order to have your personal needs met. You may never show this side of yourself in public, particularly since the 4th house is the most personal and hidden point in the chart. However, you must learn how to take time for yourself to explore your personal, emotional needs.

North Node in Cancer in the 5th House/ South Node in Capricorn in the 11th House

With your South Node in Capricorn in the 11th house, you encounter your South Node gifts of ambition, responsibility, and self-reliance when you socialize with your friends and peers. The 11th house is where we seek a sense of social and intellectual security. In the 11th house we want to belong to a social community. Within this community, we learn appropriate behavior standards, we participate in group creativity, and we learn how to receive and accept the love of others. You may find that your role in your social group is the voice of reason, and that you are the one who insists that boundaries be respected and observed. While your strength and stability may well form the backbone of your group of friends, you must be careful to avoid the trap of the South Node in Capricorn in the 11th house and become so concerned with your responsibilities to the group and with preserving the traditions and structures of the group that you are no longer able to let go and simply enjoy being with your friends. The 11th house relates to the love that we receive from others, and the trap of the South

Node in Capricorn in the 11th house tells us that to be open enough to receive love from others is to be weak and irresponsible.

The key to balancing this energy is to work with the North Node in Cancer in the 5th house. The 5th house is where we search for security in our identity as an individual. Everything that we can do that makes us feel special and unique, that makes us feel like we deserve the attention and acknowledgment of others, is found in the 5th house—that is why the 5th house is associated with such a mixed bag of concepts including children, the arts, gambling, and love affairs. With the North Node in Cancer in the 5th house, you are encouraged to focus on your own individual emotional needs rather than on your responsibilities to your peers. The focus here is more on one-to-one interactions, with you being able to form emotional bonds with other individuals, sharing and expressing your true self and your deepest needs and longings. Your South Node in Capricorn will help you maintain an awareness of good boundaries. Using those gifts as an anchor, you can begin to explore and express your emotional nature without worrying about whether you're being responsible or holding the group together.

North Node in Cancer in the 6th House/
South Node in Capricorn in the 12th House

With your South Node in Capricorn in the 12th house, you may not be consciously aware of your gifts of ambition, responsibility, and self-reliance, even though they are very evident to other people. The 12th house is like our shadow: we can't see it directly because we're usually facing the light (of consciousness, that is), but it's very visible to everyone else. Our awareness of our 12th house seems to come from our unconscious and our subconscious. You have an unconscious and intuitive need for structure and for boundaries, and an instinctive appreciation and respect for authority. And while your practical nature may be instinctive and unconscious to you, so are the traps of the South Node in Capricorn in the 12th house. You may not be aware of when you have fallen into the traps; of when you stubbornly persist in supporting the status quo, and enforcing the rules to the letter, even when the rules are clearly no longer appropriate and it's time for the status quo to shift. You may also not be aware of when you blindly submit to the demands of authority figures, regardless of how you personally feel about what you have been asked to do.

The key to balancing this energy is to work with the North Node in Cancer in the 6th house. While the 12th house relates to our emotional and soul needs, the 6th house relates to our routine physical needs. The 6th house is related to illness because if we're not taking care of our body with proper nutrition, rest, and exercise on a daily basis, we will get sick. The 6th house is also related to our job and work environment, which is an essential part of maintaining our physical existence. The North Node in Cancer in the 6th house emphasizes that you must look for the emotional connection in everything you do in your daily life. You are encouraged to take your South Node appreciation for the rules and apply that with the understanding that while the rules are meant to protect and support, that without the human element, without the emotional and caring energy, the rules are not truly effective. The 6th house is often related to our work environment and relationships with our co-workers, and with the North Node in Cancer in the 6th house, you are likely to be drawn to jobs where you can experience a nurturing energy and a family-like atmosphere with your co-workers.

North Node in Cancer in the 7th House/
South Node in Capricorn in the 1st House

With your South Node in Capricorn in the 1st house, your gifts of ambition, responsibility, and self-reliance are very much a part of your sense of self; you access them naturally and may not even be consciously aware of them. The 1st house contains everything that we identify as being a fundamental part of who we are. In fact, it's quite difficult to gain perspective on the contents of the 1st house because it's too close to us. Your respect for authority and tradition and your diligent support of boundaries and structures are second nature to you, and you may not even recognize that these skills don't come naturally to everyone. The trap of the South Node in Capricorn in the 1st house is that you may not only find it difficult to gain perspective on the *gifts* of the South Node in Capricorn, but you may also have trouble gaining perspective on the *traps* of the South Node. You may become so focused on your accomplishments, on your responsibilities, and on your need to be self-reliant that you may find it difficult to form emotional connections with other people, and even more difficult to ask for or to accept the help and support of others. The trap of the South Node in Capricorn in the 1st house tells you that you must be able to make it on your own, and that you must be able to support others, although you're not entitled to receive any support yourself.

The key to balancing this energy is to work with the North Node in Cancer in the 7th house. The 7th house is the house of one-to-one relationships, and the biggest challenge of this house is that we tend to give away the planets in this house, projecting them on others and experiencing the energies and lessons of these planets through our relationships. Working with the North Node in Cancer may be a challenge because you may feel that you lack the qualities of Cancer, and as a result, you will tend to attract people to you who embody these qualities. These relationships will help you learn how to accept and integrate the lessons of the North Node in Cancer in the 7th house. The trap of the South Node in Capricorn is to take responsibility for these people and attempt to meet their needs—something that brings out the worst in both Cancer and Capricorn energy. Rather, let these relationships show you that you can experience emotional connections; that it's not only acceptable but necessary to have others in your life on whom you can depend. You must learn that needing, giving, or receiving emotional support is not a sign of weakness, irresponsibility, or chronic dependence on others.

North Node in Cancer in the 8th House/
South Node in Capricorn in the 2nd House

With your South Node in Capricorn in the 2nd house, your gifts of ambition, responsibility, and self-reliance are important skills and resources to you. The 2nd house represents our personal resources; anything that we can call our own belongs in the 2nd house, and everything in the 2nd house helps reinforce our sense of individuality. Our possessions, resources, skills, and talents, as well as our physical body and senses, help define and support our identity as an individual. You have the ability and the experience to build up your resources and your material wealth. You are likely to take a very conservative approach to your finances, and to place great importance on your material security. The trap of the South Node in Capricorn in the 2nd house is to become too focused on the material and the physical, using your assets and wealth to protect yourself, and ultimately letting your entire sense of self-worth become tied to your accomplishments and resources on the physical plane.

The key to balancing this energy is to work with the North Node in Cancer in the 8th house. The 8th house is where we go to find a sense of emotional and soul security and self-worth. We seek a sense of inner peace through letting go of our physical and material attachments and merging with another individual on a deep, healing, and transformational level. The 8th house is also related to shared resources, and, like the 7th house, we often project our 8th-house planets on others, experiencing them through relationships. The North Node in Cancer in the 8th house encourages you to move beyond your physical accomplishments and to focus on your emotional wealth—the connections that you share with other individuals. While you can certainly benefit from sharing your physical resources with a partner, something that will force you to let go of any individual attachments you may have to those resources, the important element with the North Node in Cancer in the 8th house is to pay attention to the emotional aspects. There are needs that your physical wealth can never address—needs that can only be met through emotional connections—and these are what you are meant to explore.

North Node in Cancer in the 9th House/
South Node in Capricorn in the 3rd House

With your South Node in Capricorn in the 3rd house, your gifts of ambition, responsibility, and self-reliance will tend to manifest in your ability to communicate and experience your immediate environment. The 3rd house represents the world that is most familiar to us, and the ways that we make sense of the world through logic, reason, and language. The 3rd house also relates to our individual spiritual practices and beliefs—the ways that we connect to our spirituality outside of the structures of organized religion. You have a fundamental need to understand how your world works—what the rules are, where the boundaries are, and what is expected of you as an individual. So long as you understand these structures, you feel safe and are able to explore and express yourself and communicate with confidence, knowing that you are acting in a responsible and self-restrained manner. The trap of the South Node in Capricorn in the 3rd house is to become too dependent on these structures and rules. As we grow and evolve as individuals, our environments must grow and change with us, and this often requires changes to the structures and rules as well. When you feel threatened, you may find that you become authoritarian, insisting that things stay the way they are, if for no other reason than that's the way that things have always been.

The key to balancing this energy is to work with the North Node in Cancer in the 9th house. The 9th house is where we expand our understanding of the world and how we relate to it as an individual, and it includes philosophy, higher education, religion, and long journeys. The higher purpose that you seek through your North Node in Cancer in your 9th house has to do with your emotional connections with other individuals. As you grow and evolve, the tangible structures of the 3rd house no longer meet your needs; you find that you must move beyond your familiar world in order to feel complete on a soul level instead of simply feeling secure and protected on a mundane and physical level. The North Node in Cancer in the 9th house encourages you to explore and discover the universal truth that we all need each other; that every individual can share love and support with every other individual. Ultimately, you can learn to define your community by the emotional connections that you share, rather than the social structures that divide us.

North Node in Cancer in the 10th House/
South Node in Capricorn in the 4th House

With your South Node in Capricorn in the 4th house, your gifts of ambition, responsibility, and self-reliance are closely related to your private life and family connections. The 4th house represents our home, our families (of choice and of origin), and our connection to our past through our ancestors and tribal heritage. The 4th house is our foundation—the rock on which we build our individual life. It supports us, and through the foundation of the 4th house, we are able to achieve the public, individual accomplishments of the 10th house. You are apt to have a very strong sense of your family identity and of your core sense of self. Your family traditions, structure, and responsibilities are also likely to be quite important to you. The trap of the South Node in Capricorn in the 4th house is that you may get too caught up in your family responsibilities and in believing that you have to take care of your personal, private life that you aren't able to explore your individual path, your public life, and your career. Although you may be very attached to the structure and integrity of your family and private life, your responsibilities there may mean that you aren't able to have your own emotional needs met. You are so busy providing for others that you don't have time to think of yourself.

The key to balancing this energy is to work with the North Node in Cancer in the 10th house. The 10th house is the most public part of the chart. It relates to our career and our life path; it is where we as individuals find our most significant accomplishments. The North Node in Cancer teaches that you must learn to accept the help and emotional support of others. With your North Node in Cancer in your 10th house, you will find this support and nurturing in your public life and through your choice of career. What is important here is for you to allow yourself to experience and express your emotional identity as part of your public self. Your South Node in Capricorn in your 4th house means that you're already very grounded in the practical and that you have a fundamental respect for structure and tradition. By working with your North Node in Cancer in your 10th house, you will learn when it is appropriate for your needs as an individual to take precedence over your responsibilities to others.

North Node in Cancer in the 11th House/
South Node in Capricorn in the 5th House

With your South Node in Capricorn in the 5th house, you encounter your gifts of ambition, responsibility, and self-reliance whenever you explore and express your personal creativity. The 5th house is where we search for security in our identity as an individual. Everything that we can do that makes us feel special and unique, that makes us feel like we deserve the attention and acknowledgment of others, is found in the 5th house—that is why the 5th house is associated with such a mixed bag of concepts including children, the arts, gambling, and love affairs. You have a natural ability to give form and structure to your personal expression. You are likely to take great pride in your individual accomplishments, particularly those that come from your need to express your creativity for its own sake. The 5th house also relates to the love we give to others, and in this arena, more than any other, is where we find the trap of the South Node in Capricorn in the 5th house. Capricorn is such a self-reliant, mature, responsible, structured sign that it often has difficulty expressing love and nurturing emotions. Capricorn is very big on taking care of the material and physical needs of others, and on protecting loved ones in tangible ways, but Capricorn is not always able to get the hang of nurturing and experiencing emotional support. If you fall into the trap of the South Node in Capricorn in the 5th house, you may find that you are not able to receive love from others easily, because you are too guarded and structured in your approach to be able to give love freely.

The key to balancing this energy is to work with the North Node in Cancer in the 11th house. The 11th house is where we seek a sense of social and intellectual security. In the 11th house we want to belong to a social community. Within this community, we learn appropriate behavior standards, we participate in group creativity, and we learn how to receive and accept the love of others. With your North Node in Cancer in your 11th house, your focus is on nurturing and being nurtured by your friends and peers. This energy is about letting go of the Capricorn belief that you are an isolated individual who is responsible for the well-being of the other members in the group. Instead, working with the North Node in Cancer in the 11th house, you can experience what it feels like to truly be a part of a family, one where everyone supports and is supported by everyone else. The gifts of your South Node in Capricorn will ensure that you are always a contributing member of the group and that you give as much as you receive. But the lesson of your North Node in Cancer in your 11th house is to learn how to love, nurture, and support your friends—and to let them return the favor.

North Node in Cancer in the 12th House/
South Node in Capricorn in the 6th House

With your South Node in Capricorn in the 6th house, you encounter your gifts of ambition, responsibility, and self-reliance in your daily routines. The 6th house contains everything that we must do on a daily basis to maintain our physical existence. The 6th house is related to illness because if we're not taking care of our body with proper nutrition, rest, and exercise on a daily basis, we will get sick. The 6th house is also related to our job and work environment, which is an essential part of maintaining our physical existence. You are acutely aware of your responsibilities on a day-to-day basis. You are able to apply structure and discipline to your daily routine, and to accomplish your goals both on a personal level and in your work environment in a practical, structured, and systematic manner. The trap of the South Node in Capricorn in the 6th house, however, is that you may tend to become entirely focused on your physical and material needs and objectives; your responsibilities and commitments may begin to weigh you down and become a burden to you.

The key to balancing this energy is to work with the North Node in Cancer in the 12th house. Where the 6th house relates to how we take care of our physical health, the 12th house is how we take care of our spirit and our soul. In the 12th house we encounter our need to feel a part of a spiritual and emotional community; to let go of our individuality and to merge once again with the universe. We retreat to the 12th house whenever we need to take a break from the demands of our daily routines. With your North Node in Cancer in the 12th house, you must learn how to connect with your emotional and feeling nature; to nurture others and to allow yourself to be nurtured. Setting aside some time every day for meditation and reflection is an excellent way to start working with this energy. You can also encounter the lessons of the North Node in Cancer in the 12th house through volunteering your time and compassion and helping people who are less fortunate, particularly through working with larger institutions such as homeless shelters or elderly care facilities.

North Node Capricorn/South Node Cancer

The gifts of the South Node in Cancer include compassion, warmth, sympathy, and the ability to form and maintain nurturing, emotionally supportive connections with other individuals. You have a natural ability to nurture and to protect, and to provide a strong family structure. You bring into this life a vast resource of emotional experience and emotional sensitivity. You are able to express your emotions freely and directly, and even though you may at times appear "needy," you at least understand what your needs are and are able to ask that they be met. The trap of the South Node in Cancer, however, is the tendency to become overly dependent on the emotional level. This can manifest as being needy and always relying on the support and attention of others, or it can manifest as becoming the eternal caretaker, always worrying about taking care of other people and helping them whether they have asked for assistance or not. In either case, the key is to work with the energy and lessons of the North Node in Capricorn, which first and foremost teach self-reliance and personal responsibility.

Although you certainly want to maintain your South Node in Cancer gifts of being open and available on an emotional level, the North Node in Capricorn will help you discover how you can take responsibility for your own life and your own needs and free yourself from being dependent on the energy and attention of others for your survival. The North Node in Capricorn can also help you understand that when you are asked to support and nurture others, you must do so in a way that helps them discover their own self-reliance, rather than in ways that make them dependent on your energy for survival.

North Node in Capricorn in the 1st House/ South Node in Cancer in the 7th House

With your South Node in Cancer in the 7th house, you will encounter your South Node gifts of emotional connections, nurturing, empathy, and compassion through your relationships. One of the challenges of the 7th house, however, is that we tend to give away planets in this house, projecting them on others. You may feel that you lack the Cancer qualities, and therefore you may tend to attract people into your life who embody them. Until you are able to accept the gifts of the South Node in Cancer as being a part of you, you will experience and encounter them through your relationships. Rather than acknowledging your South Node in Cancer gifts as your own, you may feel that you lack the ability to form emotional connections and to be nurturing and caring on an individual level. You are likely to attract and to be attracted to individuals who embody the emotional nature of Cancer, so that you can experience this energy through your relationships. You must be careful to avoid the trap of the South Node in Cancer, however, and make sure that you do not form relationships based on emotional need and on your ability and desire to protect, provide for, and support your partners. These types of codependent relationships also relate to another aspect of the South Node in Cancer trap—the tendency to give and nurture others continually, out of a fear that if you stop, others will not nurture you in return, and your own needs will not be met.

Working with your North Node in Capricorn in the 1st house can help you balance this energy. However, we face a similar challenge with planets in the 1st house—while we tend to project the 7th house on others, we tend to overlook planets in the 1st house because they are such a fundamental part of who we are as an individual. With Capricorn energy in your 1st house, you are quite capable of taking responsibility for yourself, of meeting your own needs, of becoming self-reliant, and of making a contribution to society. More than that, you also understand the importance of personal boundaries and of "tough love" wherein sometimes you have to stop protecting others and force them to learn how to take responsibility for meeting their own needs. Your one-to-one relationships will keep you in touch with your emotional nature; working with your North Node in Capricorn in the 1st house will help you maintain balance in these relationships and ensure that you are well aware of the appropriate boundaries and division of responsibilities in your relationships.

North Node in Capricorn in the 2nd House/
South Node in Cancer in the 8th House

With your South Node in Cancer in the 8th house, you will encounter and experience your South Node gifts of emotional connections, nurturing, empathy, and compassion when you are experiencing close emotional connections with other individuals. The 8th house is where we go to find a sense of emotional and soul security and self-worth. We seek a sense of inner peace through letting go of our physical and material attachments and merging with another individual on a deep, healing, and transformational level. The 8th house is also related to shared resources, and, like the 7th house, we often project our 8th-house planets on others, experiencing them through relationships. But the 8th house also relates to our individual journey through our shadow self, where we seek to confront, understand, and ultimately integrate our demons. This is often a shared process, and with your South Node in Cancer in your 8th house, you have the ability to initiate emotional connections with others; and more than that, you may find that you have a tendency to depend on these connections a great deal. The trap of the South Node in Cancer in the 8th house is to become dependent on the support of others, both on an emotional level and often, since the 8th house does relate to shared resources, on a tangible and material level as well. Cancer energy can become exceedingly needy, and it can be a very short step from simply being able to enjoy and rely on the support of a partner, to feeling that you cannot survive without it.

The key to balancing this energy is to work with the North Node in Capricorn in the 2nd house. The 2nd house represents our personal resources; anything that we can call our own belongs in the 2nd house, and everything in the 2nd house helps reinforce our sense of individuality. Our possessions, resources, skills, and talents, as well as our physical body and senses, help define and support our identity as an individual. With your North Node in Capricorn in your 2nd house, one of your most important and valuable resources is your ability to take responsibility for yourself as an individual and to be self-reliant and independent. It's not necessary to cut all ties to others, of course—your ability to experience deep emotional connections is an important gift. But Capricorn's practical, focused, responsible, and self-sustaining energy can help you learn how to help yourself. When you realize that you are perfectly capable of meeting your own needs, you will be able to truly enjoy shared experiences through your South Node without feeling insecure and dependent in the process.

North Node in Capricorn in the 3rd House/
South Node in Cancer in the 9th House

With your South Node in Cancer in the 9th house, you encounter your gifts of emotional connections, nurturing, empathy, and compassion on a more philosophical and intellectual level. The 9th house is where we expand our understanding of the world and how we relate to it as an individual, and it includes philosophy, higher education, religion, and long journeys. You may tend to seek an emotional connection through 9th-house pursuits. Your beliefs and ideals are very important to you and provide great emotional support and comfort. The trap of the South Node in Cancer in the 9th house is that you may tend to become too dependent on the ideas, philosophies, and teachings of others. As you buy into one particular idea, religion, or philosophy, you may begin to lose your ability to form your own ideas and explore your own beliefs. You may become attached to a single guru or teacher, and instead of exploring and living the teachings, you may become entirely too dependent on the teacher.

The key to balancing this energy is to work with the North Node in Capricorn in the 3rd house. The 3rd house relates to our familiar environment and to how we communicate, reason, and make connections within that environment. The 3rd house also relates to our individual spiritual practices and beliefs—the ways that we connect to our spirituality outside of the structures of organized religion. The North Node in Capricorn in the 3rd house encourages you to take your beliefs and ideals and to put them into tangible use in your daily life. It's not enough that you talk the talk, you must also walk the walk. Ultimately, you are learning how to take full, personal responsibility for your beliefs, philosophies, and thoughts. You must take what you learned from the South Node in Cancer in the 9th house and actually apply it in your life. Your beliefs and ideals are meant to form the foundation of your life; use them to create and maintain your own personal code of honor and conduct. Know that the origin of the beliefs is not important: what matters is how you, as a self-reliant, responsible individual, work with them.

North Node in Capricorn in the 4th House/
South Node in Cancer in the 10th House

With your South Node in Cancer in the 10th house, you encounter your South Node gifts of emotional connections, nurturing, empathy, and compassion as a part of your public self. The 10th house is where we seek to create a tangible manifestation of our individual identity. It is where we want to make our mark on the world, and where we make our contributions to society as an individual. The 10th house is often associated with our career, but more accurately it is our life path, which is often quite different from our job. Your emotional strength, your nurturing and compassionate nature, is something that brings you great recognition and that you are able to express in a very public manner. You may find that you are drawn to careers that allow you to care for large groups of individuals, to protect children, or even to feed people, as Cancer is also very closely related to food. The trap of the South Node in Cancer in the 10th house is that all of your emotional focus and energy may be directed to your public life and your career. You may become dependent on public emotional displays and on taking care of strangers, and tend to neglect your own individual needs and your responsibilities to your family.

The key to balancing this energy is to work with the North Node in Capricorn in the 4th house. While the 10th house is the most public part of the chart, and also the most focused on individuality, the 4th house is the most private part of the chart, and the area where we are the most connected to others. The 4th house is our home; it is where we connect with our families (both of origin and of choice); but more than this, the 4th house represents our roots, our past, our ancestry, and our tribal heritage. It is the foundation on which we build our individuality and our public self. With your North Node in Capricorn in your 4th house, you must be willing to take responsibility for your roots; to help take care of your family; and to explore how your individual identity contributes to the structure, integrity, and traditions of your ancestors. This journey is quite personal, and although you will draw on the gifts of your South Node in Cancer in the 10th house and on your public persona, ultimately your goal is to learn how to bring some of the emotional connections and support that come so easily to you in your professional and public life into your private life.

North Node in Capricorn in the 5th House/
South Node in Cancer in the 11th House

With your South Node in Cancer in the 11th house, you encounter your South Node gifts of emotional connections, nurturing, empathy, and compassion when you socialize with your friends and peers. The 11th house is where we seek a sense of social and intellectual security. In the 11th house we want to belong to a social community. Within this community, we learn appropriate behavior standards, we participate in group creativity, and we learn how to receive and accept the love of others. You may find great comfort and support through your friends, and a great sense of support and community in your social groups and circles. In many ways, your friends may seem more like your true family to you than your family of origin. The trap of the South Node in Cancer in the 11th house, however, is that you may find that you tend to rely on the support and approval of your friends, or else that you tend to be a "mother hen" and be overly protective of others. One of the arenas of the 11th house is the love that we receive from others, but if you fall into the trap of the South Node in Cancer in the 11th house, you won't allow yourself to experience this love; it will either never be enough, because you have become so dependent on the support of your friends, or else you will be so focused on taking care of everyone else that you never give anyone the chance to express love for you and take care of your needs.

The key to balancing this energy is to work with the North Node in Capricorn in the 5th house. The 5th house is where we search for security in our identity as an individual. Everything that we can do that makes us feel special and unique, that makes us feel like we deserve the attention and acknowledgment of others, is found in the 5th house—that is why the 5th house is associated with such a mixed bag of concepts including children, the arts, gambling, and love affairs. The North Node in Capricorn in the 5th house teaches you that you must create a tangible manifestation of your own unique individual identity. In a sense, your lesson is to learn how to take responsibility for yourself and the things that make you uniquely you. The North

Node in Capricorn in the 5th house teaches that it is okay for you to go off on your own sometimes, and that part of learning how to be responsible for yourself is taking the time to do the things that you want to do, even if you have to do them on your own. Taking this level of responsibility for your own self-worth will prevent you from becoming too dependent on the support of your friends, and at the same time will allow you to share emotional connections with your friends in a responsible, well-structured manner.

North Node in Capricorn in the 6th House/ South Node in Cancer in the 12th House

With your South Node in Cancer in the 12th house, you may not be consciously aware of your gifts of emotional connections, nurturing, empathy, and compassion, even though they are very evident to other people. The 12th house is like our shadow: we can't see it directly because we're usually facing the light (of consciousness, that is), but it's very visible to everyone else. Our awareness of our 12th house seems to come from our unconscious and our subconscious. You have an unconscious need for emotional connections and, on some level, an unconscious fear that your core emotional needs will not be met. The 12th house relates to our spiritual connections, and the gifts of the South Node in Cancer in the 12th house enable you to experience spiritual connections on many levels, including on an emotional level. You have an instinctive understanding of the spiritual laws and truths that show that all of our needs are already met, since we are all part of all of creation. The trap of the South Node in Cancer in the 12th house is that until you learn to truly live and accept this law, you are very susceptible to the ego-based fears of Cancer. You may find that you fear your inability to meet your emotional needs, and therefore form inappropriate emotional connections with others in an attempt to meet your emotional needs.

The key to balancing this energy is to work with the North Node in Capricorn in the 6th house. While the 12th house relates to our emotional and soul needs, the 6th house relates to our routine physical needs. The 6th house is related to illness because if we're not taking care of our body with proper nutrition, rest, and exercise on a daily basis, we will get sick. The 6th house is also related to our job and work environment, which is an essential part of maintaining our physical existence. The North Node in Capricorn in the 6th house teaches that you must be willing to take responsibility for yourself, for your actions, and for meeting your individual needs on a daily basis. Because Capricorn is an earth sign, your focus will be more on the physical than on the emotional and spiritual. How you manage your physical body, your resources, and your responsibilities is directly related to how well you feel your emotional and spiritual needs are being met. Because the 6th house also relates to service that we perform for others, the 6th house is closely related to our job, our work environment, and our relationships with our co-workers. Through your work environment, you can learn both how to be self-sufficient as an individual, and also a contributing member of the group.

North Node in Capricorn in the 7th House/ South Node in Cancer in the 1st House

With your South Node in Cancer in the 1st house, your gifts of emotional connections, nurturing, empathy, and compassion are very much a part of your sense of self; you access them naturally and may not even be consciously aware of them. The 1st house contains everything that we identify as being a fundamental part of who we are. In fact, it's quite difficult to gain perspective on the contents of the 1st house because it's too close to us. You may tend to define yourself in terms of the emotional bonds that you share with others, and you interact with the world in ways that nurture and protect. The trap of the South Node in Cancer in the 1st house, however, is that you may become entirely dependent on either giving or receiving emotional support as a means of defining and expressing your individuality. The trap of the South Node in Cancer can manifest either as being too emotionally needy and dependent on others, or as being too protective and nurturing of others (and not allowing others to return the emotional support).

The key to balancing this energy is to work with the North Node in Capricorn in the 7th house. The 7th house is the house of one-to-one relationships, and the biggest challenge of the 7th house is that we tend to give away the planets in this house, projecting them on others and experiencing the energies and lessons of these planets through our relationships. Working with the North Node in Capricorn may be a challenge because you may feel that you lack the qualities of Capricorn. As a result, you will tend to attract people to you who embody these qualities. These relationships will help you learn how to accept and integrate the lessons of the North Node in Capricorn in the 7th house. Your first challenge is to learn to own your own North Node and discover your own sense of personal responsibility and self-reliance. That you will first encounter and experience these lessons through your relationships—the very place where your South Node in Cancer gifts and traps will tend to manifest in the most noticeable and significant ways—is just another part of your challenge. You may find that you tend to attract individuals into your life who seem to embody the mature, stable, responsible, and emotionally unavailable Capricorn qualities. While they are able to provide and protect you on a physical level, they will not tend to meet your emotional needs. As you learn to recognize, accept, and express the Capricorn energy of your North Node in your own life, you will be able to bring both yourself and your relationships into balance. As you learn how to be more self-reliant and responsible, you will be able to attract individuals who are simultaneously stable, secure, and emotionally available.

North Node in Capricorn in the 8th House/
South Node in Cancer in the 2nd House

With your South Node in Cancer in the 2nd house, your gifts of emotional connections, nurturing, empathy, and compassion are important skills and resources to you. The 2nd house represents our personal resources; anything that we can call our own belongs in the 2nd house, and everything in the 2nd house helps reinforce our sense of individuality. Our possessions, resources, skills, and talents, as well as our physical body and senses, help define and support our identity as an individual. Your ability to form emotional bonds with others, to nurture, and to receive support from others is fundamentally connected to your sense of self. The trap of the South Node in Cancer in the 2nd house is to become so sensitive to your emotional needs that you become overwhelmed by your needs and afraid that they will not be met. Your entire sense of self-worth can become dependent on the emotional connections that you are able to initiate and maintain with others. The more dependent you become on these connections, the more needy you will become, and the more difficult it will be to sustain these bonds because others will grow tired of the continual energetic drain. The fewer the emotional bonds, the lower your self-esteem, and the more you will believe that you are unable to meet your own survival needs without the help of others.

The key to balancing this energy is to work with the North Node in Capricorn in the 8th house. The 8th house is where we go to find a sense of emotional and soul security and self-worth. We seek a sense of inner peace through letting go of our physical and material attachments and merging with another individual on a deep, healing, and transformational level. The 8th house is also related to shared resources, and, like the 7th house, we often project our 8th-house planets on others, experiencing them through relationships. With the North Node in Capricorn in the 8th house, you are encouraged to learn how to meet your own needs through receiving the support and assistance of others. It's perfectly acceptable for you to ask for help—but what you must request is help in discovering your own boundaries and your own strengths. Rather than simply letting the Capricorn individuals whom you attract take over the management of your resources, learn from how they take responsibility in their own lives and in the ways that they are willing to assist you. In this way, you will learn to become more self-reliant. While you will be able to share emotional connections with these individuals, they will help you define appropriate boundaries for yourself, because they will not allow you to drain their energy and will not allow you to become dependent on them.

North Node in Capricorn in the 9th House/
South Node in Cancer in the 3rd House

With your South Node in Cancer in the 3rd house, your gifts of emotional connections, nurturing, empathy, and compassion will tend to manifest in your ability to communicate and experience your immediate environment. The 3rd house represents the world that is most familiar to us, and the ways that we make sense of the world through logic, reason, and language. The 3rd house also relates to our individual spiritual practices and beliefs—the ways that we connect to our spirituality outside of the structures of organized religion. You are likely to have formed many of your core beliefs and expectations of how easily your emotional needs are likely to be met at a relatively young age. While the gifts of the South Node in Cancer in the 3rd house can create a very comfortable, secure, nurturing environment for you, the trap of the South Node in Cancer in the 3rd house is to become so attached to the familiar, to creature comforts, and to having others nurture and protect you, that you are afraid to venture outside of your comfortable little world, and ultimately you avoid taking personal responsibility for your life.

The key to balancing this energy is to work with the North Node in Capricorn in the 9th house. The 9th house is where we expand our understanding of the world and how we relate to it as an individual, and it includes philosophy, higher education, religion, and long journeys. Because of the Capricorn influence in your 9th house, you will be attracted to the more structured and disciplined 9th-house pursuits, where the rules are clearly defined. Your 9th-house quest involves a search to learn about responsibility, both on an individual level and on a broader, more societal level. The North Node in Capricorn in the 9th house teaches how to take responsibility not only for your own life, but also how to contribute in a larger way, supporting the community. Higher education and religion both offer well-established paths of advancement and somewhat rigid structures, and these avenues in particular may be of interest because of the support that is offered—while you are learning how to take personal responsibility, you are also being protected by the institutions, and this can make the journey away from the familiar world of the South Node in Cancer in the 3rd house an easier one.

North Node in Capricorn in the 10th House/ South Node in Cancer in the 4th House

With your South Node in Cancer in the 4th house, your gifts of emotional connections, nurturing, empathy, and compassion are closely related to your private life and family connections. The 4th house represents our home, our families (of choice and of origin), and our connection to our past through our ancestors and tribal heritage. The 4th house is our foundation—the rock on which we build our individual life. It supports us, and through the foundation of the 4th house, we are able to achieve the public, individual accomplishments of the 10th house. Your South Node in Cancer gifts are a fundamental part of your soul identity. The emotional connections and spiritual bonds that you feel with your family are your roots and your strength. The trap of the South Node in Cancer in the 4th house, however, is to become too focused on those roots and too involved with your familial obligations, and not leave the nest and grow and advance as an individual. You must learn how to express your individual identity, and not to identify yourself solely in terms of your family.

The key to balancing this energy is to work with the North Node in Capricorn in the 10th house. The 10th house is the most public part of the chart. It relates to our career and our life path; it is where we as individuals find our most significant accomplishments. The North Node in Capricorn in the 10th house encourages you to take on public responsibility, to expand your ability to nurture and support beyond that of your family, and to be of service on a larger scale, to society. Through the North Node in Capricorn in the 10th house, you seek to create a tangible expression of your unique individual identity in a very public and prominent way. This is how you will be recognized as an individual, distinct from your family, your past, your ancestors, and your tribe. However, the only way that you will be able to accomplish this task is by drawing strength from your South Node in Cancer in the 4th house. The lesson is to learn how to use your emotional ties to your family and your past to support you and to help you excel as an individual.

North Node in Capricorn in the 11th House/ South Node in Cancer in the 5th House

With your South Node in Cancer in the 5th house, you encounter your gifts of emotional connections, nurturing, empathy, and compassion whenever you explore and express your personal creativity. The 5th house is where we search for security in our identity as an individual. Everything that we can do that makes us feel special and unique, that makes us feel like we deserve the attention and acknowledgment of others, is found in the 5th house—that is why the 5th house is associated with such a mixed bag of concepts including children, the arts, gambling, and love affairs. Your artistic and creative expressions are very much your "children." You have very strong emotional ties to your creations, and they have tremendous personal meaning to you. The trap of the South Node in Cancer in the 5th house is that you may become so personally involved in your self-expression and creativity that you become stifled and unable to grow. Your artistic "children" are much like actual children in that you can't create them alone: they require the input of more than one person. You must also be willing to share your creativity with others.

The key to balancing this energy is to work with the North Node in Capricorn in the 11th house. The 11th house is where we seek a sense of social and intellectual security. In the 11th house we want to belong to a social community. Within this community, we learn appropriate behavior standards, we participate in group creativity, and we learn how to receive and accept the love of others. With your North Node in Capricorn in the 11th house, you are encouraged to share your creative expression and your emotional connections with your friends, and to accept their help, support, and suggestions. Participating in the collaborative creative process can help take your creativity and your self-expression to the next level. The North Node in Capricorn in the 11th house will show you where you can take more responsibility and make an individual contribution to the group, and where you need to accept the help and support of others.

North Node in Capricorn in the 12th House/
South Node in Cancer in the 6th House

With your South Node in Cancer in the 6th house, you encounter your gifts of emotional connections, nurturing, empathy, and compassion in your daily routines. The 6th house contains everything that we must do on a daily basis to maintain our physical existence. The 6th house is related to illness because if we're not taking care of our body with proper nutrition, rest, and exercise on a daily basis, we will get sick. The 6th house is also related to our job and work environment, which is an essential part of maintaining our physical existence. Your co-workers are apt to feel more like family to you than merely fellow employees. Food may play heavily into your nurturing style, and you may find that when you need emotional support, food is a natural substitute. Because of this, you must be very aware of the food choices you make, as the 6th house relates to our health, and if you are not being nourished by your food, your health will tend to suffer. The trap of the South Node in Cancer in the 6th house follows in much the same way: if you become too dependent on the emotional connections, you will become afraid that your core emotional needs will not be met, and as a result your health may suffer.

The key to balancing this energy is to work with the North Node in Capricorn in the 12th house. Where the 6th house relates to how we take care of our physical health, the 12th house is how we take care of our spirit and our soul. In the 12th house we encounter our need to feel a part of a spiritual and emotional community; to let go of our individuality and to merge once again with the universe. We retreat to the 12th house whenever we need to take a break from the demands of our daily routines. With the North Node in Capricorn in the 12th house, you are encouraged to take responsibility for your own spiritual needs, to recognize that you do not have to rely on others to meet your emotional needs, and to learn how to maintain stronger emotional boundaries in your relationships. The 12th house is very personal and is best accessed through solitary pursuits, meditation, yoga, and personal spiritual practices. Because Capricorn is an earth sign, you can also tap into this energy through working with tangible things such as gardening or pottery. The more in touch you become with your North Node in Capricorn in the 12th house, the more you will recognize that, since you are a part of all of creation, all of your needs have already been met.

7

THE LEO/AQUARIUS
NODAL AXIS

The Leo/Aquarius nodal axis is the axis of group dynamics. The purpose of this axis is to learn to find a balance between prominence in the group and equality in the group. Both Leo and Aquarius are fixed signs and so are concerned with self-worth, and both signs are very much related to our ability to express and receive unconditional love. Leo expresses love from the heart on an individual basis, while Aquarius expresses love on a more abstract and mental level, offering perspective and compassion for all of humanity. True unconditional love comes when the head and the heart are balanced, a lesson that we can learn from the Leo/Aquarius nodal axis.

Leo seeks to express itself from the heart in a warm, open, generous, and creative way, and to share its unique identity with others. Leo seeks to earn its rightful place in the group by giving freely of itself, by opening the heart, loving others, and receiving love and acceptance in return. Leo does not always operate on this highest level, however, and when the ego becomes involved, Leo can become very dependent on the approval and attention of others and may tend to seek constant validation of its unique and individual identity. Leo's desire to shine and share its warmth with others can quickly become a need to be in the spotlight continually. When this happens, Leo's generosity can become conditional and motivated by the need for validation and attention, rather than coming from a place of unconditional love and acceptance.

Aquarius, on the other hand, seeks acceptance by the group based on equality. Aquarius is an entirely group-oriented energy. Rather than operating on an individual level, Aquarius identifies with the group as a separate entity. Aquarius is capable of putting the needs of the group ahead of the needs of any individuals in the group, and because of this, Aquarius energy can contribute greatly to the overall safety and quality of life of the entire group. In its highest manifestation, Aquarius energy works tirelessly to ensure that the structures and rules that protect and support the group continue to provide for the greatest degree of personal freedom within the group. Aquarius truly believes that everyone is created equal, with the same rights and privileges, and that everyone deserves the same chances to live their lives in the way in which they choose. Aquarius, however, can easily lose sight of the fact that the group is made up of individuals, each of whom is unique. While Aquarius has a great love of humanity, it often has difficulty relating to individual humans. Objectivity and perspective can be taken to the extreme, where they ultimately prevent Aquarius from expressing any true compassion.

Working with the Leo/Aquarius nodal axis teaches how to relate to humanity both as a group identity (through Aquarius) and on an individual level (through Leo). Unconditional love requires a balance between these two points of view. It must come from the heart, and is best shared on an individual basis, but it must also be tempered with perspective from the head and the understanding that everyone is equally deserving of love and acceptance.

North Node Leo/South Node Aquarius

The South Node in Aquarius brings gifts of personal freedom and tolerance of others. The South Node in Aquarius can easily identify with the greater needs of society and willingly take part in efforts to change and improve the world for the benefit of all, thanks to the South

Node's gift of selfless dedication to humanity. You are able to be impartial, objective, and absolutely fair in your judgments. You truly believe that everyone deserves equal rights and equal responsibilities. The trap of the South Node in Aquarius, however, is letting the head rule the heart. You may become too abstract and too idealistic, finding it easier to be compassionate towards strangers and groups than towards individuals and friends. Ultimately the trap is to believe that the good of the group is more important than the good of the individuals that make up that group.

Working with the lessons of the North Node in Leo will balance this energy, because the North Node teaches us to come from the heart. The contributions that Leo makes to the group are personal and unique expressions of individual identity. While Aquarius gives to the group as a whole, Leo gives to the group one individual at a time. The North Node in Leo can help the South Node in Aquarius understand that it's not just that everyone in the group is equal: it's that everyone in the group is equally special and unique—and each individual deserves to be acknowledged and appreciated for his or her contributions to the whole.

North Node in Leo in the 1st House/
South Node in Aquarius in the 7th House

With your South Node in Aquarius in the 7th house, you will encounter your South Node gifts of objectivity, personal freedom, tolerance, and humanitarianism through your relationships. One of the challenges of the 7th house, however, is that we tend to give away planets in this house, projecting them on others. You may feel that you lack the Aquarius qualities, and therefore you may tend to attract people into your life who embody them. Until you are able to accept the gifts of the South Node in Aquarius as being a part of you, you will experience and encounter them through your relationships. Because Aquarius is such a group-oriented energy, you will naturally attract people into your life who share the same ideals as you do, and who have extremely strong convictions. Your first instinct, however, may be to assume that you are simply adopting the qualities of the people you admire, rather than acknowledging that you also have these qualities. The trap of the South Node in Aquarius in the 7th house is that you may begin to define yourself as an individual through the relationships and group associations that you hold, and that since Aquarius energy holds the idea of equality above all else, you are apt to lose sight of your uniqueness and your individual creative gifts.

Working with your North Node in Leo in the 1st house can help you balance this energy and tap into your individual creative abilities. However, we face a similar challenge with planets in the 1st house— while we tend to project the 7th house on others, we tend to overlook planets in the 1st house because they are such a fundamental part of who we are as an individual. Exploring the North Node in Leo in the 1st house means exploring your own individuality. You must discover and acknowledge your personal talents, warmth, generosity, and creativity, and more importantly, you must be willing to share these qualities with others. Your South Node in Aquarius in the 7th house, and the relationships that you form around this energy, will keep you grounded and make sure that you don't upset the balance of equality in the group; but at the same time, working with the North Node in Leo in the 1st house will help you stay in touch with what you as an individual contribute to the group, and how important that contribution is.

North Node in Leo in the 2nd House/
South Node in Aquarius in the 8th House

With your South Node in Aquarius in the 8th house, you will encounter and experience your South Node gifts of objectivity, personal freedom, tolerance, and humanitarianism when you are experiencing close emotional connections with other individuals. The 8th house is where we go to find a sense of emotional and soul security and self-worth. We seek a sense of inner peace through letting go of our physical and material attachments and merging with another individual on a deep, healing, and transformational level. The 8th house is also related to shared resources, and, like the 7th house, we often project our 8th-house planets on others, experiencing them through relationships. The group-oriented energy of Aquarius expands in the 8th house to create a true sense of community, where everyone's resources are available for the entire group to access. Without putting a political spin on it, this energy is very much like the theory of communism: from each according to their abilities, and to each according to their needs. And much like communism, this works beautifully in theory, but can cause trouble in practice. Your sense of connection to a group is tied in with the amount of your personal resources that you are contributing to the group and sharing with others. The trap of the South Node in Aquarius in the 8th house is that because everyone in the group is meant to be equal, no individual contribution is given more weight or value, and this can diminish your sense of self-worth as an individual.

The key to balancing this energy is to work with the North Node in Leo in the 2nd house. The 2nd house represents our personal resources; anything that we can call our own belongs in the 2nd house, and everything in the 2nd house helps reinforce our sense of individuality. Our possessions, resources, skills, and talents, as well as our physical body and senses, help define and support our identity as an individual. The North Node in Leo in the 2nd house teaches you to value and appreciate your unique contributions to the group—even when that group consists of one-to-one relationships. You are a unique, warm, generous individual, and your contributions to any relationship or group dynamic deserve to be recognized, if only briefly, as do the contributions of every other member of the group. The North Node in Leo in the 2nd house can teach you how to ask for and receive the recognition that you deserve in life.

North Node in Leo in the 3rd House/
South Node in Aquarius in the 9th House

With your South Node in Aquarius in the 9th house, you encounter your gifts of objectivity, personal freedom, tolerance, and humanitarianism on a more philosophical and intellectual level. The 9th house is where we expand our understanding of the world and how we relate to it as an individual, and it includes philosophy, higher education, religion, and long journeys. The abstract nature of Aquarius is very much at home in the 9th house. In the 9th house we can happily spend our time living entirely in our mind, exploring the mental planes, building and dissecting theories, and entirely avoiding the need for specific application of those theories or interaction with individuals. This, of course, is the trap of the South Node in Aquarius in the 9th house. While the 9th house provides the structure and support to explore the energy and gifts of Aquarius, it also makes it very easy for Aquarius to retreat entirely into the abstract. It becomes more important to be a member of a group with shared ideals and philosophies, and less important to actually act on and live those philosophies and beliefs.

The key to balancing this energy is to work with the North Node in Leo in the 3rd house. The 3rd house relates to our familiar environment, and to how we communicate, reason, and make connections within that environment. The 3rd house also relates to our individual spiritual practices and beliefs—the ways that we connect to our spirituality outside of the structures of organized religion. The North Node in Leo in the 3rd house asks you to explore what it is that you, as an individual, truly believe. You must be willing to come down from the abstract and theoretical realm and to see how your group ideals and philosophies actually work in practice. What can you do as an individual to embody and express your beliefs? The first step, of course, is to recognize the balance between individuals and the group dynamic. It's not only that we are all equal; it's that we are all equally special. It's not that no one individual deserves the attention and acknowledgment of the group for his or her individual contributions; it's that everyone deserves it. Your contributions come by taking the abstract values of your South Node in Aquarius in the 9th house and living them, incorporating them in your own life, in your own unique way.

North Node in Leo in the 4th House/
South Node in Aquarius in the 10th House

With your South Node in Aquarius in the 10th house, you encounter your South Node gifts of objectivity, personal freedom, tolerance, and humanitarianism as a part of your public self. The 10th house is where we seek to create a tangible manifestation of our individual identity. It is where we want to make our mark on the world, and where we make our contributions to society as an individual. The 10th house is often associated with our career, but more accurately it is our life path, which is often quite different from our job. The South Node in Aquarius in the 10th house indicates that you are always ready to fight injustice, to campaign for freedom and equality, and are willing to devote yourself completely to the cause in any way that you are able. The trap of the South Node in Aquarius in the 10th house, however, is that your cru-sades can very quickly become excuses not to take responsibility for your own life and your own personal obligations. The energy and focus that you direct towards your public life, while certainly valuable, go largely unrecognized, and the more closely you identify with the group, the less in touch you are with your individuality.

The key to balancing this energy is to work with the North Node in Leo in the 4th house. While the 10th house is the most public part of the chart, and also the most focused on individuality, the 4th house is the most private part of the chart, and the area where we are the most connected to others. The 4th house is our home; it is where we connect with our families (both of origin and of choice); but more than this, the 4th house represents our roots, our past, our ancestry, and our tribal heritage. It is the foundation on which we build our in-dividuality and our public self. The North Node in Leo in the 4th house encourages you to give priority to your personal life. In your public life, you can't see how your individual efforts make a difference, but with your family, you can. By opening your heart to those closest to you, by sharing your unique warmth and love, you will be reminded of how important you are as an individual—not just as a warm body. When you share your energy and passion with your family, you are valued and appreciated immediately. And through living and sharing the ideals and beliefs of your South Node in Aquarius in the 10th house with your family, you continue to make an important contribution to your life's path—you're simply doing so on a more personal and immediate level.

North Node in Leo in the 5th House/
South Node in Aquarius in the 11th House

With your South Node in Aquarius in the 11th house, you encounter your South Node gifts of objectivity, personal freedom, tolerance, and humanitarianism when you socialize with your friends and peers. The 11th house is where we seek a sense of social and intellectual security. In the 11th house we want to belong to a social community. Within this community, we learn appropriate behavior standards, we participate in group creativity, and we learn how to receive and accept the love of others. The 11th house and Aquarius share certain concepts, most particularly a focus on the collective or group energy rather than a focus on the individual. With the South Node in Aquarius in the 11th house, you are likely to choose friends and peers who share common interests, ideals, and values, and it is this connection that forms the basis of the friendships. The trap of the South Node in Aquarius in the 11th house, however, is to become too attached to the values and ideals of the group, because ultimately this will limit your ability to express yourself as an individual. You may even find that you feel you must choose between exploring your own ideas and maintaining your friendships, because your individual expression could upset the dynamic and integrity of the group.

The key to balancing this energy is to work with the North Node in Leo in the 5th house. The 5th house is where we search for security in our identity as an individual. Everything that we can do that makes us feel special and unique, that makes us feel like we deserve the attention and acknowledgment of others, is found in the 5th house—that is why the 5th house is associated with such a mixed bag of concepts including children, the arts, gambling, and love affairs. Through working with your North Node in Leo in the 5th house, you can gain more perspective on your relationships with your friends and recognize that while your shared beliefs and ideals are a powerful bond, that it is not the *only* reason that you are friends. Your friends accept you and want you in their lives because of your unique, special gifts, your warmth, honesty, and generosity. The North Node in Leo in the 5th house can help you express yourself, with or without your friends along for the ride. You can be an individual and still maintain your relationships with your friends; they do not need to be involved in every aspect of your life. Your friends enhance your life; they do not define it.

North Node in Leo in the 6th House/
South Node in Aquarius in the 12th House

With your South Node in Aquarius in the 12th house, you may not be consciously aware of your gifts of objectivity, personal freedom, tolerance, and humanitarianism, even though they are very evident to other people. The 12th house is like our shadow: we can't see it directly because we're usually facing the light (of consciousness, that is), but it's very visible to everyone else. Our awareness of our 12th house seems to come from our unconscious and our subconscious. When you take time for yourself, you tap into your South Node gifts, and reconnect with your ideals and beliefs, letting them nourish your soul and your spirit. However, because the 12th house is difficult to access on a conscious level, the trap of the South Node in Aquarius in the 12th house is an unconscious and a subconscious drive to avoid attention and to search for a group and a cause with which you can identify. The unconscious Aquarius belief that we are all equal can mean that you tend to hide your own unique gifts and talents in an effort to fit in and be accepted by the group.

The key to balancing this energy is to work with the North Node in Leo in the 6th house. While the 12th house relates to our emotional and soul needs, the 6th house relates to our routine physical needs. The 6th house is related to illness because if we're not taking care of our body with proper nutrition, rest, and exercise on a daily basis, we will

get sick. The 6th house is also related to our job and work environment, which is an essential part of maintaining our physical existence.
With the North Node in Leo in the 6th house, you must learn how to
shine a bit every single day. The lesson here is to be able to accept that
you are worthy of praise and attention from others because of who
you are, and so is everyone else in the world. The 6th house can relate
to how we are of service to others, and what you can do is simply be
yourself and share your warmth, generosity, love, and creativity with
each person you encounter. As you recognize each person for how
special they are and how valuable their contributions are to the world,
you can also learn to accept the praise and recognition that they reflect back to you, acknowledging your contributions as well.

North Node in Leo in the 7th House/
South Node in Aquarius in the 1st House

With your South Node in Aquarius in the 1st house, your gifts of objectivity, personal freedom, tolerance, and humanitarianism are very
much a part of your sense of self; you access them naturally and may
not even be consciously aware of them. The 1st house contains everything that we identify as being a fundamental part of who we are. In
fact, it's quite difficult to gain perspective on the contents of the 1st
house because it's too close to us. You may tend to be a bit too idealistic in your approach to the world, but you are always fair in your judgments, and you are strong in your beliefs. Fighting for your rights and
freedom, as well as for the freedom of others, may be instinctive to
you. The trap of the South Node in Aquarius in the 1st house is that
you may tend to identify too much with the group, particularly when
you relate to other individuals on a one-to-one basis. Group identification can mean that you do not take personal responsibility for your
individual actions. You must be reminded to interact with people as
individuals, not as abstract concepts.

The key to balancing this energy is to work with the North Node
in Leo in the 7th house. The 7th house is the house of one-to-one relationships, and the biggest challenge of this house is that we tend to
give away the planets in this house, projecting them on others and ex-

periencing the energies and lessons of these planets through our relationships. Working with the North Node in Leo may be a challenge,
because you may feel that you lack the qualities of Leo. As a result, you
will tend to attract people to you who embody these qualities. These
relationships will help you learn how to accept and integrate the lessons
of the North Node in Leo in the 7th house. The individuals you attract
to your North Node in Leo will often fascinate, frustrate, and confound
you. They will be true individuals—they will have their own opinions,
beliefs, needs, and desires, and they will seem to defy all of the con-
ventions and restrictions of the group. Most importantly, they will demand to be noticed and acknowledged for who they are, and not for
whom they associate with. Your interactions with these individuals
will help you make the same distinctions about yourself, and allow
you to relate to others in a way that tempers your objectivity with true
warmth and compassion.

North Node in Leo in the 8th House/
South Node in Aquarius in the 2nd House

With your South Node in Aquarius in the 2nd house, your gifts of objectivity, personal freedom, tolerance, and humanitarianism are important skills and resources to you. The 2nd house represents our personal resources; anything that we can call our own belongs in the 2nd
house, and everything in the 2nd house helps reinforce our sense of individuality. Our possessions, resources, skills, and talents, as well as our
physical body and senses, help define and support our identity as an
individual. Because Aquarius is such a group-oriented sign, you may
tend to measure your level of wealth and tangible possessions against
the values and standards of your peers. This, however, can lead you
into the trap of the South Node in Aquarius in the 2nd house. Because
your connection to your peers is based on your resources, you may feel
at once driven to keep up with your peers, and also limited. If your individual resources exceed those of your peers, you may feel obligated to
spread the wealth so that everyone can enjoy the same level of comfort
and security. The Aquarius drive for equality in the group can mean
that you find it difficult to enjoy your own individual resources.

The key to balancing this energy is to work with the North Node in Leo in the 8th house. The 8th house is where we go to find a sense of emotional and soul security and self-worth. We seek a sense of inner peace through letting go of our physical and material attachments and merging with another individual on a deep, healing, and transformational level. The 8th house is also related to shared resources, and, like the 7th house, we often project our 8th-house planets on others, experiencing them through relationships. The lesson of the North Node in Leo is to learn to accept your individual talents and to recognize that you are entitled to be acknowledged, rewarded, and appreciated for them. You can experience this by choosing to share your resources with a partner—not out of any sense of obligation, but because you want to; it is an expression of your warmth and generosity, and comes from opening your heart to another. Your partners will acknowledge you for your gifts, and in turn will share theirs with you. You can learn not to feel guilty about any perceived imbalance or discrepancy between your resources and those of your partners—these will balance out in the long run. Instead, you can simply enjoy the act of sharing, and of being appreciated for it in return.

North Node in Leo in the 9th House/
South Node in Aquarius in the 3rd House

With your South Node in Aquarius in the 3rd house, your gifts of objectivity, personal freedom, tolerance, and humanitarianism will tend to manifest in your ability to communicate and experience your immediate environment. The 3rd house represents the world that is most familiar to us, and the ways that we make sense of the world through logic, reason, and language. The 3rd house also relates to our individual spiritual practices and beliefs—the ways that we connect to our spirituality outside of the structures of organized religion. With the South Node in Aquarius in the 3rd house, you may tend to perceive and experience the world as it relates to your group affiliations. The trap of the South Node in Aquarius in the 3rd house, however, is that you may become addicted to these familiar structures, and ultimately this will limit your freedom. You may tend to cling to the idea that everyone in your world should be equal, which will ultimately prevent you from growing as an individual because your growth and expansion will upset the dynamic of the group.

The key to balancing this energy is to work with the North Node in Leo in the 9th house. The 9th house is where we expand our understanding of the world and how we relate to it as an individual, and it includes philosophy, higher education, religion, and long journeys. The lesson here is to learn how to break out of your secure, safe, familiar group environment and to allow yourself to be acknowledged and recognized for your unique perspective and your individual contributions to the overall integrity and health of the community. This is not about becoming a leader per se, or about actually changing the dynamic of the group, but it is about accepting that each individual is responsible for bringing his or her own unique perspectives to the group, and being recognized for their contributions. By tapping into your South Node gifts of objectivity and compassion for humanity, you can overcome any fears you may have about exploring the uncharted territory of the 9th house.

North Node in Leo in the 10th House/
South Node in Aquarius in the 4th House

With your South Node in Aquarius in the 4th house, your gifts of objectivity, personal freedom, tolerance, and humanitarianism are closely related to your private life and family connections. The 4th house represents our home, our families (of choice and of origin), and our connection to our past through our ancestors and tribal heritage. The 4th house is our foundation—the rock on which we build our individual life. It supports us, and through the foundation of the 4th house, we are able to achieve the public, individual accomplishments of the 10th house. You may find that you tend to create home and personal situations where you feel like you are an equal and contributing member of the group. The trap of the South Node in Aquarius in the 4th house, however, is to become limited by your family connections and personal relationships. If you identify too strongly with your family, you will lose the sense of who you are as an individual.

The key to balancing this energy is to work with the North Node in Leo in the 10th house. The 10th house is the most public part of the chart. It relates to our career and our life path; it is where we as individuals find our most significant accomplishments. Simply put, the North Node in Leo in the 10th house teaches that you must learn to share your gifts and talents with the world. More than that, you must learn how to accept public recognition and attention for your contributions and skills. Leo energy in the 10th house can be a beacon of warmth and generosity. It shines and, at least briefly, it makes you a star. So long as you maintain your family connections though your South Node in Aquarius in the 4th house, you will always stay grounded and the attention and recognition won't go to your head. But through allowing yourself to be recognized, through sharing your gifts and talents on a more public scale, you grow and evolve as an individual, and at the same time your individual accomplishments contribute to the strength of your family.

North Node in Leo in the 11th House/
South Node in Aquarius in the 5th House

With your South Node in Aquarius in the 5th house, you encounter your gifts of objectivity, personal freedom, tolerance, and humanitarianism whenever you explore and express your personal creativity. The 5th house is where we search for security in our identity as an individual. Everything that we can do that makes us feel special and unique, that makes us feel like we deserve the attention and acknowledgment of others, is found in the 5th house—that is why the 5th house is associated with such a mixed bag of concepts including children, the arts, gambling, and love affairs. While the 5th house relates to our individual creative expression, Aquarius energy is completely focused on the group dynamic. Your creative expression is motivated and enhanced by your need to contribute to the group and to earn your place as a rightful member of the group. The trap of the South

Node in Aquarius in the 5th house is to lose sight of your individuality and to create things that you think will win the approval of the group, rather than things that are important to you and that express your unique self.

The key to balancing this energy is to work with the North Node in Leo in the 11th house. The 11th house is where we seek a sense of social and intellectual security. In the 11th house we want to belong to a social community. Within this community, we learn appropriate behavior standards, we participate in group creativity, and we learn how to receive and accept the love of others. The North Node in Leo in the 11th house encourages you to express your warmth, generosity, and creativity with your friends, and in turn to allow them to love you and appreciate you for it. When you accept that your friends love you for who you are and that you don't have to earn their approval, you will be able to find the proper balance for your creative expression. The things you create through your South Node in Aquarius in the 5th house will still relate to the shared ideas and ideals of the group, but they will also embody your own unique perspective.

North Node in Leo in the 12th House/
South Node in Aquarius in the 6th House

With your South Node in Aquarius in the 6th house, you encounter your gifts of objectivity, personal freedom, tolerance, and humanitarianism in your daily routines. The 6th house contains everything that we must do on a daily basis to maintain our physical existence. The 6th house is related to illness because if we're not taking care of our body with proper nutrition, rest, and exercise on a daily basis, we will get sick. The 6th house is also related to our job and work environment, which is an essential part of maintaining our physical existence. You are likely to be drawn to jobs that allow you to work with others as a team, and you may find that you have little patience for co-workers who seem to demand too much recognition for their individual efforts. The trap of the South Node in Aquarius in the 6th house, however, is to become too closely identified with your causes and with the group. You can begin to both lose your own sense of individuality, as well as a sense of others. As focused as you may be on helping humanity as a whole, if you become completely focused on the group, you will lose touch with what makes each of us human.

The key to balancing this energy is to work with the North Node in Leo in the 12th house. Where the 6th house relates to how we take care of our physical health, the 12th house is how we take care of our spirit and our soul. In the 12th house we encounter our need to feel a part of a spiritual and emotional community; to let go of our individuality and to merge once again with the universe. We retreat to the 12th house whenever we need to take a break from the demands of our daily routines. By taking time out of your daily routines, you can allow yourself to remember your own individual gifts and talents. The North Node in Leo in the 12th house doesn't need a great deal of public recognition; but it does mean that you need to allow yourself to acknowledge your accomplishments and your contributions to your South Node in Aquarius cause. Recognizing your own individuality and uniqueness will also remind you to do the same thing for others. The effectiveness and accomplishments of your team will increase dramatically when you take the time to appreciate each individual for their contribution to the greater effort.

North Node Aquarius/South Node Leo

The gifts of the South Node in Leo include the ability to give selflessly to others, to open the heart with warmth and generosity, and to embody and express love. You have a tremendous store of creativity and honesty, and can provide courage, strength, and self-assurance. You ultimately have a desire and a drive to create expressions of your unique self that will be of lasting value to others. The trap of the South Node in Leo, however, is that Leo's expressions of generosity and love can become dependent on the acknowledgment and approval of others. The South Node in Leo can express an almost pathological need to be the center of attention, and may continually seek out approval, acknowledgment, and validation from others for how special and unique it is. This trap leads to entirely self-centered behavior, pride, and frequently to childish behavior.

The way to avoid this trap is, of course, to work with the lessons of the North Node in Aquarius. Aquarius energy can help Leo see a bigger picture, to focus less on itself and more on how it can become a part of something bigger. Aquarius can also help Leo understand that becoming a member of a group does not necessitate a loss of individual identity. On the contrary, the group is made up of other individuals, each making their own unique contributions towards a common goal. Working with humanitarian causes, especially volunteer work, can be an excellent way of integrating both your South Node and your North Node by directing your creative energy towards a larger cause, one where the end goal is not individual recognition but recognition for the group.

North Node in Aquarius in the 1st House/ South Node in Leo in the 7th House

With your South Node in Leo in the 7th house, you will encounter your South Node gifts of warmth, love, generosity, and creativity through your relationships. One of the challenges of the 7th house, however, is that we tend to give away planets in this house, projecting them on others. You may feel that you lack the Leo qualities, and therefore you may tend to attract people into your life who embody them. Until you are able to accept the gifts of the South Node in Leo as being a part of you, you will experience and encounter them through your relationships. The individuals that you attract to your South Node in Leo will be generous, passionate, dynamic, and creative, but above all they will demand to be noticed and appreciated for how special they are. These individuals, however, are merely reflecting back to you the parts of yourself that you have not fully accepted or integrated—including the need for the attention and approval of others and the desire for continual recognition, which is the trap of the South Node in Leo in the 7th house. When you accept your South Node gifts, and are able to recognize that you yourself are warm, creative, and generous, you must also accept that a part of you craves attention, recognition, and validation from others.

Working with your North Node in Aquarius in the 1st house can help you balance this energy. However, we face a similar challenge with planets in the 1st house—while we tend to project the 7th house on others, we tend to overlook planets in the 1st house because they are such a fundamental part of who we are as an individual. One of the reasons that you may tend to project your South Node gifts (and traps) on others is that with Aquarius energy in your 1st house, you approach the world with the expectations that everyone is equal and that the group identity is more important than recognizing individuals. Once you stop denying your need for (and your entitlement to) some individual attention and recognition, you can find an appropriate point of balance where you can give and receive attention for your individual contributions without losing sight of the bigger picture. Finding this point of balance will also greatly improve the dynamics in your relationships, because as you come to terms with your own need for validation, and allow yourself to accept praise and attention when appropriate, you will find that the people in your life will also become less dependent on your continual attention and validation of them.

North Node in Aquarius in the 2nd House/
South Node in Leo in the 8th House

With your South Node in Leo in the 8th house, you will encounter and experience your South Node gifts of warmth, love, generosity, and creativity when you are experiencing close emotional connections with other individuals. The 8th house is where we go to find a sense of emotional and soul security and self-worth. We seek a sense of inner peace through letting go of our physical and material attachments and merging with another individual on a deep, healing, and transformational level. The 8th house is also related to shared resources, and, like the 7th house, we often project our 8th-house planets on others, experiencing them through relationships. With the South Node in Leo in the 8th house, you are able to form strong emotional bonds with other individuals, and want to open your heart to others. The trap of the South Node in Leo in the 8th house, however, is the tendency to become too dependent on the validation and approval of other people. Leo energy will do anything to be loved and appreciated, and in the 8th house you may find that you are too open and generous with your personal resources and your personal space, giving too much away to your partners in order to keep their attention focused on you.

The key to balancing this energy is to work with the North Node in Aquarius in the 2nd house. The 2nd house represents our personal resources; anything that we can call our own belongs in the 2nd house, and everything in the 2nd house helps reinforce our sense of individuality. Our possessions, resources, skills, and talents, as well as our physical body and senses, help define and support our identity as an individual. The North Node in Aquarius in the 2nd house can help you learn how to maintain strong, appropriate boundaries when it comes to your personal resources. Because Aquarius energy is primarily concerned with equality in the group, and is far less concerned with paying attention to individuals, working with the North Node in Aquarius can help you temper your need for approval, validation, and attention from others. This will help you be more objective about how you distribute and share your resources, and, more importantly, will ensure that when you do share your resources with others, that it is an equitable and fair arrangement.

North Node in Aquarius in the 3rd House/
South Node in Leo in the 9th House

With your South Node in Leo in the 9th house, you encounter your gifts of warmth, love, generosity, and creativity on a more philosophical and intellectual level. The 9th house is where we expand our understanding of the world and how we relate to it as an individual, and it includes philosophy, higher education, religion, and long journeys. The South Node in Leo in the 9th house could indicate past-life experiences where you were in a position of power and authority, and your thoughts and ideas were revered and of great importance to many other people. While your ideas and philosophies sill have great merit in this lifetime, and still carry the same creative energy, the trap of the South Node in Leo in the 9th house is that you may still crave the kind of attention and validation for your beliefs that you once enjoyed. This can make you intolerant of others when they question or debate your beliefs; because your beliefs and ideas are such a powerful expression of your individual identity, you may tend to view any dissenting opinions as a personal attack.

The key to balancing this energy is to work with the North Node in Aquarius in the 3rd house. The 3rd house relates to our familiar environment, and to how we communicate, reason, and make connections within that environment. The 3rd house also relates to our individual spiritual practices and beliefs—the ways that we connect to our spirituality outside of the structures of organized religion. The lesson and challenge here is to recognize that everyone has their own philosophy and belief system; that everyone has the right to choose their own spiritual and religious path, and that no one of these paths or beliefs has more intrinsic value than any other. Aquarius energy is about becoming part of the group collective; about finding out what you have to contribute to the group, and what it is that you share in common with the group. Your lesson is to learn how to bring your own unique beliefs and philosophies into your daily life, to be willing to share them with others, and to accept that while your contributions to the group will be appreciated, if others don't completely agree with you, it is not meant to be a personal attack.

North Node in Aquarius in the 4th House/
South Node in Leo in the 10th House

With your South Node in Leo in the 10th house, you encounter your South Node gifts of warmth, love, generosity, and creativity as a part of your public self. The 10th house is where we seek to create a tangible manifestation of our individual identity. It is where we want to make our mark on the world, and where we make our contributions to society as an individual. The 10th house is often associated with our career, but more accurately it is our life path, which is often quite different from our job. The dynamic and creative energy of Leo can easily put you in a position of prominence in your public life, where you are frequently recognized and appreciated for your unique, individual efforts. The trap of the South Node in Leo in the 10th house, however, is to put all of your energy into your career, continually seeking out more validation and more public recognition while avoiding your personal life and your family connections. Spending time with family could quickly become something to avoid, because at home all of the attention isn't on you.

The key to balancing this energy is to work with the North Node in Aquarius in the 4th house. While the 10th house is the most public part of the chart, and also the most focused on individuality, the 4th house is the most private part of the chart, and the area where we are the most connected to others. The 4th house is our home; it is where we connect with our families (both of origin and of choice); but more than this, the 4th house represents our roots, our past, our ancestry, and our tribal heritage. It is the foundation on which we build our individuality and our public self. The North Node in Aquarius in the 4th house urges you to reconnect with your family, and to allow your membership in this group to become a more prominent aspect of who you are as an individual. Even though the Aquarius energy will mean that when you are home, everyone in your family is considered to be equal, it's important to remember that this means that each member of your family is entitled to the same amount of attention and validation. Working with the North Node in Aquarius in the 4th house, however, can help you feel a part of something larger than yourself, and the security that comes from that experience is as important as individual validation.

North Node in Aquarius in the 5th House/
South Node in Leo in the 11th House

With your South Node in Leo in the 11th house, you encounter your South Node gifts of warmth, love, generosity, and creativity when you socialize with your friends and peers. The 11th house is where we seek a sense of social and intellectual security. In the 11th house we want to belong to a social community. Within this community, we learn appropriate behavior standards, we participate in group creativity, and we learn how to receive and accept the love of others. With your South Node in Leo in the 11th house, you take great pleasure in your friendships and enjoy giving of yourself to your friends. You must also be aware, however, of the trap of the South Node in Leo in the 11th house, which can lead you to base too much of your sense of self-worth on the amount of attention, love, and affection that you feel you receive from your friends. You must be careful to avoid the trap of always needing to be the center of attention in your friendships.

The key to balancing this energy is to work with the North Node in Aquarius in the 5th house. The 5th house is where we search for security in our identity as an individual. Everything that we can do that makes us feel special and unique, that makes us feel like we deserve the attention and acknowledgment of others is found in the 5th house—that is why the 5th house is associated with such a mixed bag of concepts including children, the arts, gambling, and love affairs. With the North Node in Aquarius in the 5th house, what you're really searching for is your own individual relationship to the group. Through your personal creativity and by giving of yourself, you can discover how these contributions support and enhance the group as a whole. You can discover, ultimately, why you are special and why you belong with your group of friends. Once you truly appreciate what it is that you bring to your friendships, you will become more secure in these relationships and will not require the constant validation of your friends. This will also allow you to recognize that each one of your friends is equally special and each one deserves attention and appreciation for who they are.

North Node in Aquarius in the 6th House/
South Node in Leo in the 12th House

While the South Node in Leo in the 12th house means that you are naturally warm, generous, and creative, you may not be consciously aware of your South Node gifts. Meditation and solitary spiritual practices can help you get in touch with them. The trap of the South Node in Leo in the 12th house is that not only are you unaware of your warmth, generosity, and creativity, but you are also unaware of your need for validation and appreciation from others for these qualities (although other people are very much aware of this need!). When you are feeling ignored and taken for granted, you may find that you act out to get attention, and because these needs are rooted in the unconscious of your 12th house, you may not fully understand why you behave in this way.

The key to balancing this energy is to work with the North Node in Aquarius in the 6th house. While the 12th house relates to our emotional and soul needs, the 6th house relates to our routine physical needs. The 6th house is related to illness because if we're not taking care of our body with proper nutrition, rest, and exercise on a daily basis, we will get sick. The 6th house is also related to our job and work environment, which is an essential part of maintaining our physical existence. The North Node in Aquarius in the 6th house encourages you to focus your energy on a much bigger project than your own needs. You need to be a part of a larger group, in an environment where your efforts are appreciated but also kept in perspective. You are most likely to find this through your choice of job, and you are likely to be happiest in a job where you are a part of a team and have shared goals and objectives. The sense of community that you will find through your group projects will help you avoid the traps of the South Node in Leo; you will no longer be so dependent on the validation of others, because you can clearly see how your contribution has helped accomplish the group objectives.

North Node in Aquarius in the 7th House/
South Node in Leo in the 1st House

With your South Node in Leo in the 1st house, your gifts of warmth, love, generosity, and creativity are very much a part of your sense of self; you access them naturally and may not even be consciously aware of them. The 1st house contains everything that we identify as being a fundamental part of who we are. In fact, it's quite difficult to gain perspective on the contents of the 1st house because it's too close to us. Your charisma, charm, and strength of character may seem quite unremarkable to you, and you may not, in fact, be aware of how strong your personality is. The trap of the South Node in Leo in the 1st house is that you may not be aware of the times when you are demanding too much attention from others, and when you are seeking validation for your own sense of identity and of self-worth. You may have a tendency to become very ego-involved, particularly when you are not the center of attention.

The key to balancing this energy is to work with the North Node in Aquarius in the 7th house. The 7th house is the house of one-to-one relationships, and the biggest challenge of this house is that we tend to give away the planets in this house, projecting them on others and experiencing the energies and lessons of these planets through our relationships. Working with the North Node in Aquarius may be a challenge, because you may feel that you lack the qualities of Aquarius. As a result, you will tend to attract people to you who embody these qualities. These relationships will help you learn how to accept and integrate the lessons of the North Node in Aquarius in the 7th house. You may find that you tend to attract individuals into your life who embody the humanitarian, idealistic, and above all egalitarian focus of Aquarius energy. While these individuals are perfectly willing to give you your due when it is appropriate, they will not allow you to take up residence in the spotlight. More importantly, these individuals will be interested in causes that are bigger than one person. You can learn that sometimes being a part of something greater than any one individual is more important than having your individual contributions recognized and acknowledged.

North Node in Aquarius in the 8th House/ South Node in Leo in the 2nd House

With your South Node in Leo in the 2nd house, your gifts of warmth, love, generosity, and creativity are important skills and resources to you. The 2nd house represents our personal resources; anything that we can call our own belongs in the 2nd house, and everything in the 2nd house helps reinforce our sense of individuality. Our possessions, resources, skills, and talents, as well as our physical body and senses, help define and support our identity as an individual. You tend to enjoy your resources, and you particularly enjoy sharing them with others. When you share your wealth and your prosperity, you are opening your heart. The trap of the South Node in Leo in the 2nd house, however, is the tendency to have your sense of self-worth as an individual tied to your material assets. Rather than seeking direct validation and attention from others, you seek it through your acquisitions and possessions, and through your bank balance. The trap of the South Node in Leo in the 2nd house can also cause you to become too reckless with your resources, believing that the more you share with others, the more they will love and appreciate you.

The key to balancing this energy is to work with the North Node in Aquarius in the 8th house. The 8th house is where we go to find a sense of emotional and soul security and self-worth. We seek a sense of inner peace through letting go of our physical and material attachments and merging with another individual on a deep, healing, and transformational level. The 8th house is also related to shared resources, and, like the 7th house, we often project our 8th-house planets on others, experiencing them through relationships. The North Node in Aquarius in the 8th house will help you avoid the South Node traps, because the North Node in Aquarius can't be bought off; it's not the least bit impressed by your assets or your bank balance. The relationships that you experience through your North Node in Aquarius in the 8th house are about a meeting of the minds. The connections you experience are based on shared values and beliefs and ideals. When you work with your North Node in Aquarius in the 8th house, you will free yourself from your material attachments, because you will remember that your possessions do not define who you are as an individual—and more importantly, they don't make you any more or less special or deserving of love.

North Node in Aquarius in the 9th House/
South Node in Leo in the 3rd House

With your South Node in Leo in the 3rd house, your gifts of warmth, love, generosity, and creativity will tend to manifest in your ability to communicate and experience your immediate environment. The 3rd house represents the world that is most familiar to us, and the ways that we make sense of the world through logic, reason, and language. The 3rd house also relates to our individual spiritual practices and beliefs—the ways that we connect to our spirituality outside of the structures of organized religion. With Leo in your 3rd house, you may tend to perceive the world primarily as it relates to you. You do genuinely care about the people and objects in your personal space, and your warmth and love is very evident to them. The trap of the South Node in Leo in the 3rd house, however, is that you may tend to become too attached to your familiar, comfortable environment. Being the "big fish in a small pond" can be an addictive experience, and the idea of exploring a situation where you may not be the star, the hero, and the center of attention may be very unappealing to you.

The key to balancing this energy is to work with the North Node in Aquarius in the 9th house. The 9th house is where we expand our understanding of the world and how we relate to it as an individual, and it includes philosophy, higher education, religion, and long journeys. Working with the North Node in Aquarius in the 9th house will allow you to break out of the familiar but limiting world of the 3rd house and allow you to learn about the greater and more important role that you can play as a member of the group. Part of the challenge is to learn that becoming an equal member of the group, that sharing in a larger philosophy and participating in a larger goal, does not mean that you must give up your unique identity. You will give up the belief that the world revolves around you, but you will also give up the *need* to feel that you are the center of the universe. The North Node in Aquarius in the 9th house can teach you how to give of yourself freely and fully, without needing the approval and validation of others, and to learn how to find security and support for your individual identity through sharing in the collective dream.

North Node in Aquarius in the 10th House/ South Node in Leo in the 4th House

With your South Node in Leo in the 4th house, your gifts of warmth, love, generosity, and creativity are closely related to your private life and family connections. The 4th house represents our home, our families (of choice and of origin), and our connection to our past through our ancestors and tribal heritage. The 4th house is our foundation— the rock on which we build our individual life. It supports us, and through the foundation of the 4th house, we are able to achieve the public, individual accomplishments of the 10th house. The warmth and love that you share with your loved ones and experience in your home environment is a source of great strength to you. The trap of the South Node in Leo in the 4th house, however, is to become too attached to the attention and recognition that you receive from your loved ones and to be afraid to leave your private world behind in case you don't receive the same kind of validation from the world at large.

The key to balancing this energy is to work with the North Node in Aquarius in the 10th house. The 10th house is the most public part of the chart. It relates to our career and our life path; it is where we as individuals find our most significant accomplishments. The North Node in Aquarius in the 10th house, however, is less concerned with being recognized for your public contributions, and more concerned with your participation. You must be willing to share your warmth and creativity with the world on a larger scale, and to join with others to work towards the betterment of all of society. The North Node in Aquarius in the 10th house sees the bigger picture, and encourages you to grow as an individual, and to explore your individuality through your group connections and more universal goals and objectives. The love and attention that you receive from your family is what gives you the strength and security to follow the direction of your North Node.

North Node in Aquarius in the 11th House/
South Node in Leo in the 5th House

With your South Node in Leo in the 5th house, you encounter your gifts of warmth, love, generosity, and creativity whenever you explore and express your personal creativity. The 5th house is where we search for security in our identity as an individual. Everything that we can do that makes us feel special and unique, that makes us feel like we deserve the attention and acknowledgment of others, is found in the 5th house—that is why the 5th house is associated with such a mixed bag of concepts including children, the arts, gambling, and love affairs. You truly enjoy being creative and sharing yourself with others. Having fun is a way of expressing your unique self. The trap of the South Node in Leo in the 5th house, however, is to become addicted to having fun. You may feel that you always have to be the life of the party and the center of attention. Your dependence on the approval and validation of others may become so strong that you even begin to lose sight of your true self, and instead try to act, behave, and pretend to be whatever you think other people want you to be—anything to continue receiving their love and attention.

The key to balancing this energy is to work with the North Node in Aquarius in the 11th house. The 11th house is where we seek a sense of social and intellectual security. In the 11th house we want to belong to a social community. Within this community, we learn appropriate behavior standards, we participate in group creativity, and we learn how to receive and accept the love of others. The more you are able to focus on group activities and on projects that require the skills and energy of many people, the more you will get in touch with your North Node in Aquarius in the 11th house. Working with your friends and peers, with the people who choose to include you in their lives, can help you keep your need for approval and validation in check.

You don't need to keep proving yourself to these people—they already know that you're special. The more you are willing to acknowledge others for their abilities, the more you will be appreciated for yours.

North Node in Aquarius in the 12th House/ South Node in Leo in the 6th House

With your South Node in Leo in the 6th house, you encounter your gifts of warmth, love, generosity, and creativity in your daily routines. The 6th house contains everything that we must do on a daily basis to maintain our physical existence. The 6th house is related to illness because if we're not taking care of our body with proper nutrition, rest, and exercise on a daily basis, we will get sick. The 6th house is also related to our job and work environment, which is an essential part of maintaining our physical existence. You are likely to enjoy the attention and the acknowledgment of your co-workers, and your charm and self-confidence will certainly serve you well and help you gain advancement in the material world. The trap of the South Node in Leo in the 6th house, however, is to become too dependent on the attention you receive from others, and too focused on obtaining favor and influence in your daily life and job environment. Leo energy is associated with rulers, and rulers often choose to manipulate others and play with politics in order to maintain their control. You must be careful that your daily routines don't begin to revolve around making sure that you receive your due from everyone.

The key to balancing this energy is to work with the North Node in Aquarius in the 12th house. Where the 6th house relates to how we take care of our physical health, the 12th house is how we take care of our spirit and our soul. In the 12th house we encounter our need to feel a part of a spiritual and emotional community; to let go of our individuality and to merge once again with the universe. We retreat to the 12th house whenever we need to take a break from the demands of our daily routines. The humanitarian and group activities that you choose to pursue through your North Node in Aquarius in the 12th house may tend to be completely removed from your day-to-day life.

While it's important that you learn to devote some of your time and energy towards helping others with no thought to personal gain or recognition, your choice of a cause is likely to be a very personal one, motivated by your individual spiritual needs. The perspective and the spiritual satisfaction that you will find by working with your North Node will keep your ego in check and help you avoid the South Node traps, greatly improving the relationships in your work environment.

8

THE VIRGO/PISCES
NODAL AXIS

The Virgo/Pisces nodal axis is the axis of matter and spirit. The purpose of this axis is to learn how to balance between discrimination and assimilation, and between isolation and integration. Virgo and Pisces are both mutable signs and are concerned with healing and completion, and ultimately with perfection and improvement. Virgo operates on the physical plane, while Pisces operates on the spiritual plane. While we are incarnated on Earth, we must continually work to find the point of balance between our physical world and our true spiritual identity. We must discover how to bring spirit into matter and matter into spirit.

Virgo is concerned with perfecting the physical universe through competence, analysis, discrimination, and adaptability. Virgo excels at all forms of quality control, thanks to outstanding analytical capabilities, an attention to detail, and ultimately the genuine desire to serve the greater good of humanity through being of service to others. Virgo seeks to adapt both itself and the world in its never-ending search for perfection. Virgo must learn, however, that perfection is as much about the process as it is about the end result. Virgo has a tendency to become overly critical of both itself and others. An obsession with details can result in a loss of perspective that makes any real progress and improvements difficult. Virgo's dedication to service must also be balanced with a strong sense of individual identity, otherwise it may tend to make needless sacrifices, believing that they are required to be truly of service.

Pisces, on the other hand, seeks the perfection and healing of the spirit that will occur when we once again become one with the cosmic consciousness and all of creation. Pisces seeks the higher spiritual truth that we are all one, and operates by forming connections, dissolving boundaries, and transmuting all negative emotions and pain. By helping free others of these fear-based obstacles, Pisces encourages others to advance along their own spiritual paths. Pisces, however, can become so obsessed with the spiritual and emotional realms that it pays little attention to the physical world and, as a consequence, finds it increasingly difficult to function in a responsible manner. Pisces can also fall into the ego trap of the martyr or the victim. Instead of releasing the pain and negativity it absorbs from others, Pisces may tend to hold on to it, mistakenly believing that its personal suffering will help others evolve spiritually.

The Virgo/Pisces nodal axis teaches us how to balance our physical world with our spiritual one. Each realm is equally important, and ultimately, through working with the Virgo/Pisces axis, we can begin to understand how they are in fact a part of each other. We can learn to spiritualize our daily lives and routines, and at the same time we can learn to anchor and ground the spiritual energies in our lives.

North Node Virgo/South Node Pisces

The South Node in Pisces brings with it the gifts of compassion, spirituality, and the ability to heal and transmute negative emotions and pain. The South Node in Pisces indicates a fundamental understanding and appreciation of the ways in which we are all connected to each other. The trap of the South Node in Pisces, however, is to use spirituality as an escape and a way to avoid the lessons and responsibilities that come with being incarnated in a physical body. When the South Node in Pisces denies the physical, it begins to take on the role of the victim and the martyr, and the more suffering it experiences on the physical plane, the more determined it will become to try to escape through mystical experiences and spirituality.

The North Node in Virgo teaches that we must learn to bring spirit into matter. Rather than attempting to escape the physical, we must instead learn how to perfect and improve the physical world by applying the South Node's gifts of compassion, spirituality, and healing in practical, tangible ways in daily life. The most valuable skill that the North Node in Virgo can teach is discrimination. Pisces tends to take on the pain and suffering of others, often without being conscious of it. As we learn how to analyze and discriminate more, we can begin to separate our own issues from the pain and negativity that we have absorbed from others in the course of healing them.

North Node in Virgo in the 1st House/
South Node in Pisces in the 7th House

With your South Node in Pisces in the 7th house, you will encounter your South Node gifts of compassion, healing, and spirituality through your relationships. One of the challenges of the 7th house, however, is that we tend to give away planets in this house, projecting them on others. You may feel that you lack the Pisces qualities, and therefore you may tend to attract people into your life who embody them. Until you are able to accept the gifts of the South Node in Pisces as being a part of you, you will experience and encounter them through your relationships. Your relationships hold great potential for spiritual connections and emotional bonds. The trap of the South Node in Pisces in the 7th house, however, is that you may lose a sense of appropriate energetic and personal boundaries in your relationships, as Pisces attempts to unite you with the rest of the universe. You may, in fact, not even understand or be aware of when your boundaries are being violated by others because Pisces energy simply has no understanding of boundaries in the first place.

Working with your North Node in Virgo in the 1st house can help you balance this energy. However, we face a similar challenge with planets in the 1st house—while we tend to project the 7th house on others, we tend to overlook planets in the 1st house because they are such a fundamental part of who we are as an individual. You must learn to apply the North Node lessons of discrimination and focus to your sense of self and your understanding of who you are as an individual. As you begin to analyze and discriminate, you will gain a much clearer and more defined sense of self; in other words, you will begin to discover the appropriate boundaries that will help you know where you end and others begin. This awareness can help you contain the spiritual and energetic connections that you share in your relationships. It won't diminish the quality of the bonds that you share, it will simply ensure that you are maintaining a clear sense of self in your relationships.

North Node in Virgo in the 2nd House/
South Node in Pisces in the 8th House

With your South Node in Pisces in the 8th house, you will encounter and experience your South Node gifts of compassion, healing, and spirituality when you are experiencing close emotional connections with other individuals. The 8th house is where we go to find a sense of emotional and soul security and self-worth. We seek a sense of inner peace through letting go of our physical and material attachments and merging with another individual on a deep, healing, and transformational level. The 8th house is also related to shared resources, and, like the 7th house, we often project our 8th-house planets on others, experiencing them through relationships. With the South Node in Pisces in the 8th house, you are able to form and share close emotional connections with others, and are very comfortable with merging your resources with a partner's resources. The trap of the South Node in Pisces in the 8th house is that you may have difficulty respecting or even recognizing boundaries in your relationships. This can result in you feeling that other people aren't respecting your resources or possessions, or in you being the one who crosses other people's boundaries, using resources that they have not explicitly invited you to share.

The key to balancing this energy is to work with the North Node in Virgo in the 2nd house. The 2nd house represents our personal resources; anything that we can call our own belongs in the 2nd house, and everything in the 2nd house helps reinforce our sense of individuality. Our possessions, resources, skills, and talents, as well as our physical body and senses, help define and support our identity as an individual. Working with the North Node in Virgo in the 2nd house, you can begin to discriminate, analyze, and understand the boundaries of your personal resources. You can learn to turn a critical eye to your world and separate what is yours from what belongs to other people. From this point, you can also set clearer boundaries and define the extent to which your resources are being shared with others.

North Node in Virgo in the 3rd House/
South Node in Pisces in the 9th House

With your South Node in Pisces in the 9th house, you encounter your gifts of compassion, healing, and spirituality on a more philosophical and intellectual level. The 9th house is where we expand our understanding of the world and how we relate to it as an individual, and it includes philosophy, higher education, religion, and long journeys. The South Node in Pisces in the 9th house gives you an especially open mind as you absorb and explore the different ideas and beliefs that are held by others. The trap of the South Node in Pisces in the 9th house, however, is that you may tend to become too credulous, believing everything you hear, and that you may drift from one new concept to the next without truly experiencing or understanding them. This can also make you particularly vulnerable to people who use religion and spirituality as a way of controlling and manipulating others.

The key to balancing this energy is to work with the North Node in Virgo in the 3rd house. The 3rd house relates to our familiar environment, and to how we communicate, reason, and make connections within that environment. The 3rd house also relates to our individual spiritual practices and beliefs—the ways that we connect to our spirituality outside of the structures of organized religion. The North Node in Virgo in the 3rd house teaches you that you must take what you have experienced through your South Node in Pisces in the 9th house and put it to the test, bringing it into your daily life in a practical and tangible way. You must be willing to inspect each new belief and idea and to analyze it, and ultimately to decide if it does have any validity or use to you on a personal level. Spirituality is not something that is meant to exist outside of ourselves, nor is it something that requires us to travel great distances or to accept whatever the priests and gurus tell us. Your South Node in Pisces in the 9th house gives you a very strong personal connection to the universe, and is the source of your spiritual connection. The North Node in Virgo in the 3rd house reminds you that your objective is to apply your spiritual connections to your everyday life and your familiar environment. It's not necessary to travel to the 9th house to find your spirituality: it's always with you, in every aspect of your life.

North Node in Virgo in the 4th House/
South Node in Pisces in the 10th House

With your South Node in Pisces in the 10th house, you encounter your South Node gifts of compassion, healing, and spirituality as a part of your public self. The 10th house is where we seek to create a tangible manifestation of our individual identity. It is where we want to make our mark on the world, and where we make our contributions to society as an individual. The 10th house is often associated with our career, but more accurately it is our life path, which is often quite different from our job. Pisces energy in the 10th house announces to the world that you are someone who will heal, counsel, and make others feel better. Whether or not your choice of career specifically involves healing, you are the one that others turn to for guidance and comfort. The trap of the South Node in Pisces in the 10th house, however, is that you may get drawn into the martyr energy of Pisces and begin to lose a sense of your true individual self. Boundary issues are always a concern with Pisces, and the fact that you find it difficult to turn away anyone in pain only makes the situation worse. The more you deny your own needs and boundaries, the more your life becomes about suffering for others, which is neither the purpose of your life nor a true expression of Pisces energy.

The key to balancing this energy is to work with the North Node in Virgo in the 4th house. While the 10th house is the most public part of the chart, and also the most focused on individuality, the 4th house is the most private part of the chart, and the area where we are the most connected to others. The 4th house is our home; it is where we connect with our families (both of origin and of choice); but more than this, the 4th house represents our roots, our past, our ancestry, and our tribal heritage. It is the foundation on which we build our individuality and our public self. The North Node in Virgo in the 4th house encourages you to explore and analyze your past, your family, and your heritage to help you better understand yourself as an individual. Putting more focus and attention on your home and personal life will help you create stronger boundaries. As your sense of individuality grows, you can draw strength from your family and your heritage. This, in turn, will help you share and enjoy your South Node gifts: you will better understand how you can heal others without causing yourself any suffering in the process.

North Node in Virgo in the 5th House/
South Node in Pisces in the 11th House

With your South Node in Pisces in the 11th house, you encounter your South Node gifts of compassion, healing, and spirituality when you socialize with your friends and peers. The 11th house is where we seek a sense of social and intellectual security. In the 11th house we want to belong to a social community. Within this community, we learn appropriate behavior standards, we participate in group creativity, and we learn how to receive and accept the love of others. Your friends are apt to feel more like family to you, and you are likely to be the spiritual center of your circle of friends. The trap of the South Node in Pisces in the 11th house, however, is to merge too much with others to the extent that you lose sight of yourself and your individual contributions to the group. The less self-aware you are, the more likely it is that you will take on more of your friends' negativity and pain; and while this will make them feel better, you and your sense of self-worth may begin to suffer for it.

The key to balancing this energy is to work with the North Node in Virgo in the 5th house. The 5th house is where we search for security in our identity as an individual. Everything that we can do that makes us feel special and unique, that makes us feel like we deserve the attention and acknowledgment of others, is found in the 5th house—that is why the 5th house is associated with such a mixed bag of concepts including children, the arts, gambling, and love affairs. The North Node in Virgo in the 5th house encourages you to discover exactly what it is about you that makes you special and unique. You must be willing to turn an analytical and critical eye on yourself, and to separate the things that you share with others from the things that come from you and you alone. As you define these special, creative qualities, you will also begin to improve and perfect them. Maintaining a sense of your unique gifts will ensure that you are able to maintain good boundaries in your relationships with your friends.

North Node in Virgo in the 6th House/
South Node in Pisces in the 12th House

With your South Node in Pisces in the 12th house, you may not be consciously aware of your gifts of compassion, healing, and spirituality, even though they are very evident to other people. The 12th house is like our shadow: we can't see it directly because we're usually facing the light (of consciousness, that is), but it's very visible to everyone else. Our awareness of our 12th house seems to come from our unconscious and our subconscious. With the South Node in Pisces in the 12th house, your spiritual connections are always with you, unconsciously guiding your choices and path in life. You instinctively understand how all of creation is connected. The trap of the South Node in Pisces in the 12th house, however, is that this continuous connection to the spiritual realm often makes it difficult for you to function in the physical. Your sense of boundaries, both physical and personal, may be particularly weak. The more you connect with this spiritual longing, the less happy you will be with a physical existence.

The key to balancing this energy is to work with the North Node in Virgo in the 6th house. While the 12th house relates to our emotional and soul needs, the 6th house relates to our routine physical needs. The 6th house is related to illness because if we're not taking care of our body with proper nutrition, rest, and exercise on a daily basis, we will get sick. The 6th house is also related to our job and work environment, which is an essential part of maintaining our physical existence. Much as you might like to escape the limitations of the physical, the lesson of your North Node in Virgo in the 6th house is to learn how to operate in the physical world. You must take the spiritual awareness that you receive through your South Node and incorporate it into the daily routines that you must follow in order to maintain your physical body. The more you are willing to work with your North Node, the more comfortable you will become with the physical plane. Ultimately, you can reach the point of balance where you realize that spirit is indeed everywhere, and that you can continue to experience your soul connections to the universe even while you are experiencing (and thriving in) the material world.

North Node in Virgo in the 7th House/
South Node in Pisces in the 1st House

With your South Node in Pisces in the 1st house, your gifts of compassion, healing, and spirituality are very much a part of your sense of self; you access them naturally and may not even be consciously aware of them. The 1st house contains everything that we identify as being a fundamental part of who we are. In fact, it's quite difficult to gain perspective on the contents of the 1st house because it's too close to us. You experience and approach the world with open arms, always recognizing the true spiritual nature of your perceptions. The trap of the South Node in Pisces in the 1st house, however, is that in your eagerness to experience and connect with the world in general (and other individuals in particular), you lose a sense of your own individuality. You may not be able to determine where you end and where someone else begins. Because of your poor energetic boundaries, you instinctively take on a tremendous amount of other people's negativity. If you are not consciously aware of this and take steps to keep your energy field clear, the pain and suffering that you experience may lead you to want to escape your body, or at least your physical reality.

The key to balancing this energy is to work with the North Node in Virgo in the 7th house. The 7th house is the house of one-to-one relationships, and the biggest challenge of the 7th house is that we tend to give away the planets in this house, projecting them on others and experiencing the energies and lessons of these planets through our relationships. Working with the North Node in Virgo may be a challenge, because you may feel that you lack the qualities of Virgo. As a result, you will tend to attract people to you who embody these qualities. These relationships will help you learn how to accept and integrate the lessons of the North Node in Virgo in the 7th house. You may find that you tend to attract people into your life who are grounded, practical, analytical, and precise—all of the things, in fact, that you feel you are unable to be yourself. As you interact with these individuals, your lesson is to recognize that these qualities are within you. You can learn to discriminate and to separate your energy and your emotions from the energy and emotions of other people, which you have simply absorbed along the way. These relationships will also help you gain a clear sense of your own personal boundaries, because the Virgo individuals that you attract are *very* clear about their own boundaries and will make you aware of when you cross a line. As you learn to recognize where other people begin, this will also define where you end.

North Node in Virgo in the 8th House/
South Node in Pisces in the 2nd House

With your South Node in Pisces in the 2nd house, your gifts of compassion, healing, and spirituality are important skills and resources to you. The 2nd house represents our personal resources; anything that we can call our own belongs in the 2nd house, and everything in the 2nd house helps reinforce our sense of individuality. Our possessions, resources, skills, and talents, as well as our physical body and senses, help define and support our identity as an individual. You have the ability to see beyond the illusions of the physical and to connect and operate on the more eternal and enduring spiritual and soul level. As the 2nd house relates to your finances and possessions, this also means that you are unlikely to become too attached to the physical world. The trap of the South Node in Pisces in the 2nd house, however, is that you may tend to put such little value and attention towards your more tangible resources that you have difficulty maintaining them. It's one thing not to identify with the physical; it's quite another to deny it completely and suffer lack and insolvency because of the entirely incorrect belief that in order to be spiritual, we must deny the physical.

The key to balancing this energy is to work with the North Node in Virgo in the 8th house. The 8th house is where we go to find a sense of emotional and soul security and self-worth. We seek a sense of inner peace through letting go of our physical and material attachments and merging with another individual on a deep, healing, and transformational level. The 8th house is also related to shared resources, and, like the 7th house, we often project our 8th-house planets on others, experiencing them through relationships. The North Node in Virgo in the 8th house means that you can look to other people to assist you in maintaining your resources. Through your relationships, and the shared resources and values that you experience in your relationships, you can find the much needed perspective that will help you balance the energy of your South Node. Allowing a partner to manage your resources and observing how your partners are able to honor and respect (and enjoy) the physical can be valuable lessons for you. Ultimately, you can learn how to manage your physical resources and assets without becoming attached to them.

North Node in Virgo in the 9th House/
South Node in Pisces in the 3rd House

With your South Node in Pisces in the 3rd house, your gifts of compassion, healing, and spirituality will tend to manifest in your ability to communicate and experience your immediate environment. The 3rd house represents the world that is most familiar to us, and the ways that we make sense of the world through logic, reason, and language. The 3rd house also relates to our individual spiritual practices and beliefs—the ways that we connect to our spirituality outside of the structures of organized religion. You are likely to feel very connected to your environment, and may find that you are able to rely on your intuition and feelings to guide you when you communicate with others. The trap of the South Node in Pisces in the 3rd house is that you may also tend to accept whatever information you are given, particularly when it comes to your spiritual practices. You have the sense that you are a part of something greater than yourself, but in order to truly explore and understand that, you must be willing to leave your familiar routine behind and take a closer look at what it is that you actually do believe.

The key to balancing this energy is to work with the North Node in Virgo in the 9th house. The 9th house is where we expand our understanding of the world and how we relate to it as an individual, and it includes philosophy, higher education, religion, and long journeys. The North Node in Virgo in the 9th house encourages you to explore your spirituality in a structured manner, using analysis and focus to understand the underlying structures, philosophies, and beliefs. You are looking for an intellectual appreciation of your personal spirituality, and ultimately are seeking a balance and understanding that will enable you to clearly define your personal beliefs and to discover to what extent those beliefs are shared by others. As you explore with your North Node in Virgo in the 9th house, you will find the particular ideas and philosophies that resonate most with your personal spiritual experiences; and through analyzing and dissecting these philosophies, you will feel an even greater sense of connection to the universe.

North Node in Virgo in the 10th House/
South Node in Pisces in the 4th House

With your South Node in Pisces in the 4th house, your gifts of compassion, healing, and spirituality are closely related to your private life and family connections. The 4th house represents our home, our families (of choice and of origin), and our connection to our past through our ancestors and tribal heritage. The 4th house is our foundation—the rock on which we build our individual life. It supports us, and through the foundation of the 4th house, we are able to achieve the public, individual accomplishments of the 10th house. You are likely to share very close emotional and psychic connections with your family, and you may be very sensitive to any emotional disturbances in your environment. The trap of the South Node in Pisces in the 4th house is that you may not feel that you have a particularly solid foundation on which to build your life. Your spiritual and emotional connections to your family and your ancestors are so strong that you may find it difficult to break free and explore your individuality; more than this, you may feel so comfortable and connected that you don't want to break free.

The key to balancing this energy is to work with the North Node in Virgo in the 10th house. The 10th house is the most public part of the chart. It relates to our career and our life path; it is where we as individuals find our most significant accomplishments. Your lesson is to draw strength from your family and your spiritual connections, and to share your gifts and talents with the world by using them in a practical and tangible manner. With the North Node in Virgo in the 10th house, your life path involves learning how to discriminate, to analyze, to improve, and ultimately to perfect the physical world. Your South Node gifts of compassion and healing and your ability to draw strength from your family and your soul connections are an important part of this path. You must learn how to understand and analyze your gifts, and strive to improve your skills. You can be of great service to others in your life, but in order to do so, you must understand how to apply your spiritual gifts in a practical and tangible way.

North Node in Virgo in the 11th House/
South Node in Pisces in the 5th House

With your South Node in Pisces in the 5th house, you encounter your gifts of compassion, healing, and spirituality whenever you explore and express your personal creativity. The 5th house is where we search for security in our identity as an individual. Everything that we can do that makes us feel special and unique, that makes us feel like we deserve the attention and acknowledgment of others, is found in the 5th house—that is why the 5th house is associated with such a mixed bag of concepts including children, the arts, gambling, and love affairs. Your creativity is closely linked with your spirituality, and you may find that you have a fertile and active imagination that allows you to create things of great beauty. The trap of the South Node in Pisces in the 5th house, however, is the tendency to want to escape the demands of the physical world through fantasy, creativity, love affairs, or even spiritual pursuits.

The key to balancing this energy is to work with the North Node in Virgo in the 11th house. The 11th house is where we seek a sense of social and intellectual security. In the 11th house we want to belong to a social community. Within this community, we learn appropriate behavior standards, we participate in group creativity, and we learn how to receive and accept the love of others. With the North Node in Virgo in the 11th house, you can rely on your friends to keep you in touch with reality. As the 11th house relates to shared creativity, the analytical and discriminating elements of Virgo may be experienced through your relationships with others. The feedback and suggestions you get from your friends can provide valuable insight and help you improve the things that you create with your South Node in Pisces in the 5th house. Through your interactions with your friends, you will be able to give form and structure to your personal creative expressions, and ultimately to be able to share your compassion and your personal spirituality with others.

North Node in Virgo in the 12th House/
South Node in Pisces in the 6th House

With your South Node in Pisces in the 6th house, you encounter your gifts of compassion, healing, and spirituality in your daily routines. The 6th house contains everything that we must do on a daily basis to maintain our physical existence. The 6th house is related to illness because if we're not taking care of our body with proper nutrition, rest, and exercise on a daily basis, we will get sick. The 6th house is also related to our job and work environment, which is an essential part of maintaining our physical existence. The South Node in Pisces in the 6th house can allow you to see the true connections between everything, and to carry with you an ongoing appreciation for the perfection and wholeness of all of creation. However, the trap of the South Node in Pisces in the 6th house is that this continual state of heightened spiritual awareness may overwhelm you. In the extreme, it can make functioning in the material world on a day-to-day basis very challenging for you.

The key to balancing this energy is to work with the North Node in Virgo in the 12th house. Where the 6th house relates to how we take care of our physical health, the 12th house is how we take care of our spirit and our soul. In the 12th house we encounter our need to feel a part of a spiritual and emotional community; to let go of our individuality and to merge once again with the universe. We retreat to the 12th house whenever we need to take a break from the demands of our daily routines. The lesson here is to learn to combine your inborn understanding that every individual is a part of all of creation and shares a fundamental spiritual connection, with the ability to maintain an awareness of the (illusion of) separation of each individual. This may seem contradictory, but all that it requires is that you develop a dual awareness that allows you to express your gifts of compassion and healing while maintaining (and strengthening) your perception of boundaries, both on a physical and an energetic level.

North Node Pisces/South Node Virgo

The South Node in Virgo brings with it the gifts of a sharp, analytical mind, the ability to discriminate, to isolate, and to evaluate, and an appreciation for the rewards of service, of being able to contribute something of value to society. The South Node in Virgo has a tremendous understanding of the material and physical plane, and on many levels has mastered the art of perfecting and improving the world for the betterment of everyone. The trap of the South Node in Virgo, however, is the tendency to focus on the physical plane exclusively. The South Node in Virgo can become obsessed with details and become overly critical. It can also become entirely too dependent on the left-brain, logical, deductive approach to the world. The service that the South Node in Virgo performs is entirely in the physical and material plane, and the more focused the South Node in Virgo becomes on proving its competence and skill, the more empty it will ultimately feel.

The North Node in Pisces teaches that we must learn to see the world in terms of how connected everything is as well as how different things are. Where Virgo discriminates, Pisces integrates. More than that, the North Node in Pisces balances the left-brain functions of the South Node in Virgo by opening the door to a higher spiritual connection. The North Node in Pisces can teach compassion and forgiveness, and go a long way towards softening the critical approach of the South Node in Virgo. But most of all, the North Node in Pisces can help us understand that there must always be a higher spiritual purpose to our service: that everything we accomplish on the physical plane is simply an aspect of our work on the spiritual plane.

North Node in Pisces in the 1st House/
South Node in Virgo in the 7th House

With your South Node in Virgo in the 7th house, you will encounter your South Node gifts of discrimination, analysis, and service through your relationships. One of the challenges of the 7th house, however, is that we tend to give away planets in this house, projecting them on others. You may feel that you lack the Virgo qualities, and therefore you may tend to attract people into your life who embody them. Until you are able to accept the gifts of the South Node in Virgo as being a part of you, you will experience and encounter them through your relationships. You may find that you attract individuals who are very intellectual, logical, and left-brained. While Virgo wants to serve and to improve, Virgo can also become overly critical, and this is the trap of the South Node in Virgo in the 7th house. You may tend to be too critical of others, particularly in the context of your relationships. You want perfection in your relationships, but you may fall into the trap of focusing on what's wrong in the relationship, or worse, what's wrong with your partner. And since everything in the 7th house is really a projection and reflection of our own issues, what is actually happening is that you are being overly critical of yourself.

Working with your North Node in Pisces in the 1st house can help you balance this energy. However, we face a similar challenge with planets in the 1st house—while we tend to project the 7th house on others, we tend to overlook planets in the 1st house because they are such a fundamental part of who we are as an individual. The lesson of the North Node in Pisces in the 1st house is self-acceptance. You must learn how to accept and integrate all of the parts of yourself, both those that you like and those that you don't like. While your South Node in Virgo strives for perfection, your North Node in Pisces will teach you to rise above the illusions of the material plane and to view yourself and your life from a higher, spiritual perspective where you will realize that you are already perfect. Ultimately, everything you encounter, especially everything you encounter in your relationships, is a part of you that you must simply learn to accept, acknowledge, and integrate.

North Node in Pisces in the 2nd House/
South Node in Virgo in the 8th House

With your South Node in Virgo in the 8th house, you will encounter and experience your South Node gifts of discrimination, analysis, and service when you are experiencing close emotional connections with other individuals. The 8th house is where we go to find a sense of emotional and soul security and self-worth. We seek a sense of inner peace through letting go of our physical and material attachments and merging with another individual on a deep, healing, and transformational level. The 8th house is also related to shared resources, and, like the 7th house, we often project our 8th-house planets on others, experiencing them through relationships. With your South Node in Virgo in the 8th house, you naturally seek to perfect and improve your emotional connections with others. With respect to finances and shared resources, you are very sensitive to the details and may in fact prove to be quite adept at managing other people's resources. The trap of the South Node in Virgo in the 8th house, however, is to become too concerned with perfection, which often manifests as being overly critical—something that can be devastating in relationships. It's one thing to understand how the relationship can be improved, but it's quite another to focus on the imperfections. An additional trap of the South Node in Virgo in the 8th house is the tendency to give up your own personal needs in order to be of service to others. For example, while you may be skillful at managing other people's resources, you may end up neglecting your own.

The key to balancing this energy is to work with the North Node in Pisces in the 2nd house. The 2nd house represents our personal resources; anything that we can call our own belongs in the 2nd house, and everything in the 2nd house helps reinforce our sense of individuality. Our possessions, resources, skills, and talents, as well as our physical body and senses, help define and support our identity as an individual. The North Node in Pisces in the 2nd house challenges you to explore your own inner, spiritual world and discover that your true resources are boundless. When your South Node in Virgo in the 8th house becomes too critical of others, you are actually judging yourself for your self-perceived shortcomings. Working with the North Node in Pisces in the 2nd house will help you regain your sense of perspective and remember that on a spiritual level, you are perfect and complete. This will greatly enhance your sense of self-worth, and will allow you to be more forgiving of both yourself and others.

North Node in Pisces in the 3rd House/
South Node in Virgo in the 9th House

With your South Node in Virgo in the 9th house, you encounter your gifts of discrimination, analysis, and service on a more philosophical and intellectual level. The 9th house is where we expand our understanding of the world and how we relate to it as an individual, and it includes philosophy, higher education, religion, and long journeys. Virgo energy seeks to analyze, perfect, and improve. In the 9th house your South Node gifts can encourage you to search for your own truth by exploring and dissecting different philosophies and belief systems. The trap of the South Node in Virgo in the 9th house, however, is that you may tend to become too involved in the details and mechanics of your higher-mind activities. You may become so focused on the minute details of the theories of how you choose to live your life that you never actually find the time to apply these theories and actually experience life.

The key to balancing this energy is to work with the North Node in Pisces in the 3rd house. The 3rd house relates to our familiar environment, and to how we communicate, reason, and make connections within that environment. The 3rd house also relates to our individual spiritual practices and beliefs—the ways that we connect to our spirituality outside of the structures of organized religion. The lesson and the challenge of the North Node in Pisces in the 3rd house is to choose a philosophy (or even the best elements of a number of different philosophies) and actually experience it, incorporating it in your day-to-day life so that it can actually open you to the more spiritual and healing aspects of the universe. When you are willing to take a somewhat broader view of your ideas and experiences, and stop focusing exclusively on the details, you can experience a more integrated and intuitive approach to the world and discover a far deeper and more satisfying understanding of yourself. Part of the lesson of the

North Node in Pisces in the 3rd house is to learn to trust your instincts; to know that you are being guided, and that you will naturally and easily discover ways to make use of the information and knowledge that you gained through your South Node in Virgo in the 9th house.

North Node in Pisces in the 4th House/
South Node in Virgo in the 10th House

With your South Node in Virgo in the 10th house, you encounter your South Node gifts of discrimination, analysis, and service as a part of your public self. The 10th house is where we seek to create a tangible manifestation of our individual identity. It is where we want to make our mark on the world, and where we make our contributions to society as an individual. The 10th house is often associated with our career, but more accurately it is our life path, which is often quite different from our job. Your attention to detail, your dedication to service, and your fundamental drive to achieve perfection on the physical plane are the tools that will help you make your mark on the world. The trap of the South Node in Virgo in the 10th house, however, is to focus on your career and your public accomplishments exclusively, neglecting your personal, familial, and spiritual responsibilities.

The key to balancing this energy is to work with the North Node in Pisces in the 4th house. While the 10th house is the most public part of the chart, and also the most focused on individuality, the 4th house is the most private part of the chart, and the area where we are the most connected to others. The 4th house is our home; it is where we connect with our families (both of origin and of choice); but more than this, the 4th house represents our roots, our past, our ancestry, and our tribal heritage. It is the foundation on which we build our individuality and our public self. With the Pisces influence in your 4th house, you can connect with your roots very easily, because Pisces tends to overlook all boundaries and to remind us that we're truly connected to the entire universe. Applying your South Node gifts of analysis and discrimination to exploring and understanding your heritage and your past will both enhance your sense of individual identity and strengthen the emotional and spiritual bonds that you share with your family. You must learn to integrate your public self with your private self. Your individual accomplishments and experiences enrich and expand the tapestry of your heritage; you strive to succeed and to improve not only for your own benefit, bur also for the benefit of your ancestors and your descendants.

North Node in Pisces in the 5th House/
South Node in Virgo in the 11th House

With your South Node in Virgo in the 11th house, you encounter your South Node gifts of discrimination, analysis, and service when you so-cialize with your friends and peers. The 11th house is where we seek a sense of social and intellectual security. In the 11th house we want to belong to a social community. Within this community, we learn ap-propriate behavior standards, we participate in group creativity, and we learn how to receive and accept the love of others. You may find that being of service to your friends and peers, and appearing to be very competent in all that you do, is very important to you. You gen-uinely want to improve the quality of life for your friends, and partic-ularly when involved in group activities, you stay focused on the de-tails, setting a high standard for both yourself and the rest of the group. The trap of the South Node in Virgo in the 11th house, how-ever, is that you may become overly critical of yourself and of your friends because of your drive for perfection. You may also begin to lose your sense of individuality, and rather than participating in the group as an equal member, you may feel that you must earn your right to be included in the group by being of service or of use to them.

The key to balancing this energy is to work with the North Node in Pisces in the 5th house. The 5th house is where we search for secu-rity in our identity as an individual. Everything that we can do that makes us feel special and unique, that makes us feel like we deserve the attention and acknowledgment of others, is found in the 5th house—that is why the 5th house is associated with such a mixed bag of concepts including children, the arts, gambling, and love affairs. With your North Node in Pisces in the 5th house, you must learn to find your own personal expression of your spirituality. You may choose to share this with others, or you may choose to express it in private. Through this spiritual exploration, you will discover your own unique gifts and talents. When you are ready to create something tangible with these talents, your South Node gifts will help you take care of the details. The most important discovery you can make through your North Node in Pisces in your 5th house is that anything you create that is an expression of yourself is perfect just the way it is.

North Node in Pisces in the 6th House/
South Node in Virgo in the 12th House

With your South Node in Virgo in the 12th house, you may not be consciously aware of your gifts of discrimination, analysis, and service, even though they are very evident to other people. The 12th house is like our shadow: we can't see it directly because we're usually facing the light (of consciousness, that is), but it's very visible to everyone else. Our awareness of our 12th house seems to come from our unconscious and our subconscious. So while you may instinctively be able to focus on the details and always have ideas on how things can be improved, you may also not be aware of when you have fallen into the trap of the South Node in Virgo in the 12th house and are being overly critical of everything and everyone that you encounter. Your instinctive and unconscious desire for perfection is a powerful force that you must learn to moderate.

The key to balancing this energy is to work with the North Node in Pisces in the 6th house. While the 12th house relates to our emotional and soul needs, the 6th house relates to our routine physical needs. The 6th house is related to illness because if we're not taking care of our body with proper nutrition, rest, and exercise on a daily basis, we will get sick. The 6th house is also related to our job and work environment, which is an essential part of maintaining our physical existence. The lesson of the North Node in Pisces in the 6th house is to learn to see the perfection in everything in your life. The more you can become aware of the underlying spiritual connections that encompass every aspect of your life, the less driven you will be to try to find your spiritual strength through improving the physical world. Your South Node in Virgo desire to be of service to others can operate quite comfortably in your 6th-house environment. All you need do is stay in touch with the North Node in Pisces in the 6th house and recognize that one of the most important ways that you can be of service is to be accepting, loving, and compassionate to everyone you encounter, honoring their inherent perfection and supporting them as they work towards realizing this potential in themselves.

North Node in Pisces in the 7th House/
South Node in Virgo in the 1st House

With your South Node in Virgo in the 1st house, your gifts of discrimination, analysis, and service are very much a part of your sense of self; you access them naturally and may not even be consciously aware of them. The 1st house contains everything that we identify as being a fundamental part of who we are. In fact, it's quite difficult to gain perspective on the contents of the 1st house because it's too close to us. You are someone who truly wants to make a difference in the world; to be of use and to improve the quality of life for everyone. The trap of the South Node in Virgo in the 1st house, however, is that you are also apt to look for perfection; in your efforts to improve the world, you can be perceived as being too critical, overly analytical, and generally dissatisfied. It won't matter much to those who feel criticized by you that you are more critical of yourself than you could ever be of anyone else.

The key to balancing this energy is to work with the North Node in Pisces in the 7th house. The 7th house is the house of one-to-one relationships, and the biggest challenge of the 7th house is that we tend to give away the planets in this house, projecting them on others and experiencing the energies and lessons of these planets through our relationships. Working with the North Node in Pisces may be a challenge, because you may feel that you lack the qualities of Pisces. As a result, you will tend to attract people to you who embody these qualities. These relationships will help you learn how to accept and integrate the lessons of the North Node in Pisces in the 7th house. Until you learn to integrate these lessons, you will tend to attract people who will reflect them back to you. On the one hand, the serenity, compassion, and spirituality of these individuals will be very attractive to you. On the other hand, the fact that they aren't nearly as concerned as you are with the details of the world may be quite frustrating. The most important thing you can learn through these relationships is that these individuals accept you for who you are—they don't try to change you, and they don't criticize; they understand that you are perfect as you are. Eventually, you will be able to accept this about yourself.

North Node in Pisces in the 8th House/
South Node in Virgo in the 2nd House

With your South Node in Virgo in the 2nd house, your gifts of discrimination, analysis, and service are important skills and resources to you. The 2nd house represents our personal resources; anything that we can call our own belongs in the 2nd house, and everything in the 2nd house helps reinforce our sense of individuality. Our possessions, resources, skills, and talents, as well as our physical body and senses, help define and support our identity as an individual. Having Virgo energy in the 2nd house often indicates that you are very capable when it comes to managing and tracking your finances and material assets. The trap of the South Node in Virgo in the 2nd house, however, is to become too attached to your physical assets so that your sense of worth as an individual becomes tied to your skill in acquiring and managing your net worth. Your obsession with your resources can also become an obstacle in your relationships, because you may tend to keep score, measuring how much others appreciate you based on how they value and appreciate your money and your possessions.

The key to balancing this energy is to work with the North Node in Pisces in the 8th house. The 8th house is where we go to find a

The Virgo/Pisces Nodal Axis   203

sense of emotional and soul security and self-worth. We seek a sense of inner peace through letting go of our physical and material attachments and merging with another individual on a deep, healing, and transformational level. The 8th house is also related to shared resources, and, like the 7th house, we often project our 8th-house planets on others, experiencing them through relationships. The North Node in Pisces in the 8th house encourages you to move beyond the idea of tangible resources, and instead to seek out emotional and soul connections with others. The security you experience through your relationships and through your emotional bonds will help you keep your material connections in their proper perspective. While it's important to be responsible for your physical resources, people and relationships are more important than things.

North Node in Pisces in the 9th House/
South Node in Virgo in the 3rd House

With your South Node in Virgo in the 3rd house, your gifts of discrimination, analysis, and service will tend to manifest in your ability to communicate and experience your immediate environment. The 3rd house represents the world that is most familiar to us, and the ways that we make sense of the world through logic, reason, and language. The 3rd house also relates to our individual spiritual practices and beliefs—the ways that we connect to our spirituality outside of the structures of organized religion. You like to be of use in your immediate environment, and pride yourself on a thorough understanding of everything in your personal sphere of influence. The trap of the South Node in Virgo in the 3rd house is the drive to make everything in your personal environment absolutely perfect. The more obsessed you become with achieving this goal, the more you try to improve, analyze, and understand, the more dissatisfied (and the more critical) you will become. You will keep dissecting the details, but ultimately you're looking in the wrong direction; in order to find the satisfaction you seek, you will need to look beyond your familiar world.

The key to balancing this energy is to work with the North Node in Pisces in the 9th house. The 9th house is where we expand our understanding of the world and how we relate to it as an individual, and it includes philosophy, higher education, religion, and long journeys. In order for you to find the perfection you seek in your personal world, you will need to leave the 3rd house and work with the North Node in Pisces in the 9th house. You need to find a sense of your connection with the rest of the universe—where exactly does your personal sphere of influence fit into the greater scheme of things? You have a sense that even your smallest actions may be of great significance, but until you are able to leave the logical, analytical, and left-brained 3rd house and Virgo South Node behind, you won't begin to discover why. The North Node in Pisces in the 9th house gives you a greater perspective on the universe. As you explore new philosophies and beliefs, you will begin to view your personal world in a very different light. The more you open your mind to the spiritual perspectives, the more you will be able to keep your obsession with perfection in check.

North Node in Pisces in the 10th House/
South Node in Virgo in the 4th House

With your South Node in Virgo in the 4th house, your gifts of discrimination, analysis, and service are closely related to your private life and family connections. The 4th house represents our home, our families (of choice and of origin), and our connection to our past through our ancestors and tribal heritage. The 4th house is our foundation—the rock on which we build our individual life. It supports us, and through the foundation of the 4th house, we are able to achieve the public, individual accomplishments of the 10th house. Your analytical ability, and your desire to improve and to be of use are closely connected to your roots; the details of your family life, and being of service to your family, are very important to you. The trap of the South Node in Virgo in the 4th house, however, is that your devotion and service to your family may prevent you from exploring your greater role in the world.

The key to balancing this energy is to work with the North Node in Pisces in the 10th house. The 10th house is the most public part of the chart. It relates to our career and our life path; it is where we as individuals find our most significant accomplishments. The Pisces influence in your 10th house will help you see the connections between the experiences you have as an individual in the world, and your role as a member of your family unit. The skills and understanding of your family relationships can be used in a more public manner, allowing you to share your gifts with the world. You are encouraged to explore your creativity and your spirituality; you will always be guided by your roots, by your family, and by the things that you "know" to be true. But you must also take the time to explore the world beyond the material and the tangible; to pursue your own "spirit quest" so that you can help improve the spiritual aspects of your family as well as the physical and material aspects of it.

North Node in Pisces in the 11th House/
South Node in Virgo in the 5th House

With your South Node in Virgo in the 5th house, you encounter your gifts of discrimination, analysis, and service whenever you explore and express your personal creativity. The 5th house is where we search for security in our identity as an individual. Everything that we can do that makes us feel special and unique, that makes us feel like we deserve the attention and acknowledgment of others, is found in the 5th house—that is why the 5th house is associated with such a mixed bag of concepts including children, the arts, gambling, and love affairs. With Virgo in your 5th house, you will tend to treat the act of creation with great respect. You may find that you are driven to constantly improve your skills and techniques so that your creations truly express your artistic vision. The trap of the South Node in Virgo in the 5th house, however, is to be too critical of your creations. When your skills and artistry are not able to match your vision, you may become very frustrated. If you will settle for nothing less than perfection, then the entire purpose of the creative process is lost. And since what you want to create is an expression of your own unique identity, and your creations aren't good enough, then you will fall deeper into the pit of self-criticism and self-censorship.

The key to balancing this energy is to work with the North Node in Pisces in the 11th house. The 11th house is where we seek a sense of social and intellectual security. In the 11th house we want to belong to a social community. Within this community, we learn appropriate behavior standards, we participate in group creativity, and we learn how to receive and accept the love of others. The love and support that you experience with your friends is the key to keeping your creativity flowing. With the North Node in Pisces in the 11th house, your friends are going to love and appreciate the beauty in everything. When you share in the creative process with your friends, you can be reminded of the sheer joy in the act of creation. This can help you keep your South Node in Virgo in the 5th house in check. While aesthetics are certainly to be admired, what is most important about the things you create is that they contain a part of you, and therefore they are perfect just as they are.

North Node in Pisces in the 12th House/
South Node in Virgo in the 6th House

With your South Node in Virgo in the 6th house, you encounter your gifts of discrimination, analysis, and service in your daily routines. The 6th house contains everything that we must do on a daily basis to maintain our physical existence. The 6th house is related to illness because if we're not taking care of our body with proper nutrition, rest, and exercise on a daily basis, we will get sick. The 6th house is also related to our job and work environment, which is an essential part of maintaining our physical existence. You naturally pay attention to the details in your daily affairs; you are also likely to be attracted to the type of work that requires your analytical and discriminatory skills. The trap of the South Node in Virgo in the 6th house, however, is that you may become addicted to your routine and obsessed with the little details, which will cause you to lose perspective on your life. Routine and procedure are quite useful in their proper place, but when they become too important, the result is a soulless bureaucracy where filling out the correct forms becomes more important than actually helping others.

The key to balancing this energy is to work with the North Node in Pisces in the 12th house. Where the 6th house relates to how we take care of our physical health, the 12th house is how we take care of our spirit and our soul. In the 12th house we encounter our need to feel a part of a spiritual and emotional community; to let go of our individuality and to merge once again with the universe. We retreat to the 12th house whenever we need to take a break from the demands of our daily routines. The lessons of the North Node in Pisces in the 12th house will help you keep your perspective, to recognize that as important as the physical and material plane is, our emotional and spiritual needs are equally as important. The North Node in Pisces in the 12th house teaches compassion and forgiveness—both of yourself and of others. It can also help you remember that in order to truly be of use to others, you must be able to address more than their physical needs; service takes place on both the physical *and* the spiritual planes, simultaneously. You must learn how to incorporate your spirituality into your daily routines to keep your desire to be of service (to yourself and to others) in the proper perspective.

9

ASPECTS AND TRANSITS

Aspects to the Nodes

Aspects are angular relationships between two planets that represent a connection, a flow, and an exchange of energy between those two planets. When two planets aspect each other, each planet is affected by the other, and the relationship between the two planets changes how each planet expresses itself. The nodes, however, are not physical bodies, and this means that they do not *make* aspects; they only *receive* them. This is an important distinction: when a planet aspects the nodal axis, this has no effect on the planet at all (with one possible exception, which we'll look at shortly). So what is the difference between *making* an aspect and *receiving* an aspect? If you have Mars conjunct the North Node (and, of course, simultaneously opposing the South Node), this influences how you experience and work with your nodes because they *receive* the aspect from Mars. Mars, however, only *makes* the aspect to the nodes and therefore isn't affected by the aspect in the least. If Mars were conjunct Jupiter, on the other hand, both Mars and Jupiter would be influenced.

This also means that the nodes cannot be included when defining aspect patterns. Aspect patterns are groups of three or more *planets* that are linked by common aspects. When you hear people speak of *grand trines* and *t-squares*, these are aspect patterns. What makes an aspect pattern, however, is the fact that because all three planets are so closely connected to each other, they tend to act together—and most importantly, whenever one of the planets is activated by a transit, the other two (or more) planets are also being triggered at the same time. For more information on aspect patterns, see chapter 11 of my book *Astrology: Understanding the Birth Chart.*

Any time a planet aspects the nodal axis, it actually forms two complementary aspects, one to each node. In other words, every aspect to the nodes is either a conjunction (0°) and opposition (180°), a trine (120°) and sextile (60°), a semisextile (30°) and quincunx (150°), a semisquare (45°) and sesquiquadrate (135°), or two squares (90° each). However, the only aspects to the nodes that are worth considering are the conjunction/opposition and the squares.

When interpreting the nodes, it's essential to keep them in perspective with the rest of the chart. While the nodes relate to our spiritual and soul lessons that we are working on in this lifetime, these lessons do not necessarily operate in an obvious, tangible, or even observable manner. We live in a physical world, and while we are having our human experiences, our primary focus is on exploring the realm of the physical plane. The nodes are not physical bodies, and therefore they do not operate on the physical plane. The planets, on the other hand, most certainly do exist in the physical, and this is why the planets are always the most important things to consider in a natal chart. If we choose to become aware of our spiritual lessons, and we make the conscious choice and effort to tune in to our nodes, we can certainly become more aware of how these lessons are playing out in our lives—but we must always remember that the way we actually learn these lessons is through dealing with the physical. Becoming attuned to our spiritual compass can give us a sense of direction, but it doesn't change the fact that the path we're following is still most strongly influenced by the planets. Because of this, the subtle effects that, for example, a trine or sextile from Jupiter to the nodes may have are hardly even worth considering.

Conjunctions and Oppositions to the Nodes

Having planets conjunct one of the nodes (and opposing the other node) is an important consideration, however. Even simply having planets in the same *sign* as the nodes is significant. When a planet is conjunct one of the nodes, we are going to be in contact with that node and its related gifts, traps, and lessons every time we experience and express the energy of that planet. We are simply more familiar with the energy of a sign when we happen to have a planet in that sign in our chart, particularly when that planet is one of the "personal" planets (Sun, Moon, Mercury, Venus, Mars, Jupiter, and Saturn). These planets represent parts of ourselves that we experience and express every day; and when they're in the same sign as the nodes, this means that we're naturally going to be far more experienced in the gifts and lessons associated with the nodal axis.

When evaluating how we work with the nodal axis, it's important to consider planets that are conjunct one of the nodes. When we have planets in the same sign as one of the nodes, but not the other, we're going to be far more familiar with the energy of the conjoined node, and may find it more difficult to address and experience the energy of the opposing node. Individuals with the South Node in Cancer who also have the Moon, Mars, and Jupiter in Cancer are going to be very familiar with the Cancer gifts; they are also going to be more susceptible, of course, to the traps of the South Node in Cancer. If they have no planets in Capricorn, then the lessons of the North Node in Capricorn, and the energy of Capricorn in general, is going to be very unfamiliar to them—it's going to seem rather alien, and they will tend to experience it through relationships, regardless of whether or not their nodes fall in the 1st/7th or 2nd/8th houses. In cases like this, we must keep the lessons and objectives of the North Node in their proper perspective. With three planets in Cancer and no planets in Capricorn, the focus is heavily weighted towards working with the Cancer energy and the associated gifts of the South Node. Working with the South Node energy is going to take precedence, and the North Node in Capricorn becomes primarily a point of reference that can provide the necessary perspective to avoid the traps of the South Node in Cancer. The challenge here is not so much to learn how to embody and express the qualities of the Capricorn North Node, but rather to learn how to use the Cancer energy (and the planets in Cancer) in a more balanced and skillful manner.

This can work both ways, of course. Consider individuals with the North Node in Aries, along with the Sun, Mercury, and Venus in Aries, and only the South Node in Libra. Here we have a situation where the North Node will receive the greatest natural focus and attention. With the Sun, Mercury, and Venus in Aries, these people are unlikely to worry about giving up their individual identity in relationships! What they may tend to forget, however, is that they must maintain an awareness of their Libra South Node gifts so that they can keep harmony in their relationships. The challenge with planets conjunct the North Node is that we must remember not to abandon our South Node gifts—they provide the anchor and the foundation that allow us to explore the lessons of the North Node.

Squares to the Nodes

Planets that square the nodal axis are also important, but for a very different reason. In classical astrology, a planet that is square the nodes was said to be "at the bendings," and this planet would become a focal point of change and crisis for the individual. Dr. J. Lee Lehman discusses the history of this phenomenon in great detail in her excellent book *Classical Astrology for Modern Living* (Atglen, Pa.: Whitford Press, 1996). When interpreting a natal chart, be aware of planets that are "at the bendings" and squaring the nodes. These planets may not seem particularly prominent in their own right; if the planet is an outer planet (such as Uranus, Neptune, or Pluto), you may even be tempted to overlook its significance, barring strong aspects to personal planets. However, a planet that is "at the bendings" must be given careful consideration, because that planet, and the issues associated with it, will tend to take center stage repeatedly in the individual's life. Working on a conscious level with the lessons and gifts of the nodal axis can help bring the energy of this planet back into its proper perspective.

Transits to the Nodes

When considering planets transiting the nodes, again we only pay attention to the conjunction, squares, and opposition to the nodes. When a transiting planet makes any of these aspects to the nodes, it has the same type of effect as a natal aspect to the nodes does, albeit only for a very short period of time. As planets conjoin our South Node, these planets help us connect to the gifts of the South Node in ways that relate to the nature of the planet. Transiting Jupiter conjunct the South Node will encourage us to enjoy our South Node gifts, to explore them, and to discover new ways of working with them; transiting Saturn conjunct the South Node, on the other hand, will tend to give more structure and restraint to our use of the South Node gifts. The same applies to conjunctions to the North Node: tuning in to the planet conjoining the North Node can help us further our understanding of how to incorporate the North Node lessons into our lives.

Eclipse Cycles

The nodes make a complete cycle through the zodiac every eighteen years. The transiting nodes, however, will be in the same signs as your natal nodes every nine years. To clarify, let's assume that your North Node is in Aries and your South Node is in Libra in your birth chart. When you were nine years old, the transiting North Node was in Libra, and the transiting South Node was in Aries. This is the halfway point of the eighteen-year cycle, but the transiting nodes are once again on the Aries/Libra axis. When you were eighteen, the transiting North Node was once again in Aries, and the transiting South Node was in Libra, completing the old cycle and starting a new one.

Eclipses create change and crisis in our life; the lessons that we face as a result of the eclipses are very much related to our soul lessons. This is particularly the case when the eclipses occur in the same signs and houses as your nodal axis. Every nine years we are forced to address lessons and tests that specifically relate to the lessons of our nodal axis. As is always the case with the nodes, we may not always be able to make the connection between the events that we must resolve in the physical and the lessons that our nodes seem to present to us. However, making a conscious effort to balance the gifts of your South Node with the lessons of your North Node can help you weather the challenges brought on by the eclipses.

APPENDIX:
"ASK KEVIN"

The articles that follow are drawn from the "Ask Kevin" section of my website (www.astro-horoscopes.com). "Ask Kevin" is designed to be a place where people can submit their questions about astrological technique and interpretation. The topics covered here can rarely be found in general books on astrology. I've selected a few questions from the "Ask Kevin" archives that specifically relate to the nodes and to eclipses.

Annette Writes:

"What does it mean if you are born during a Full Moon and a total lunar eclipse? Does this denote special meaning?"

Kevin Answers:

Annette,

Thanks for your question! Let me start out by clarifying some of the astronomy involved and then we'll take a look at the astrology. Lunar eclipses can only occur at a Full Moon. They happen whenever the Full Moon occurs within 12 degrees of the Moon's nodes. The Earth's shadow is cast over the Moon, blocking out the Moon's light for a period of time.

During a lunar eclipse, the Sun is conjunct one node, and the Moon is conjunct the other node. In a natal chart, this means that the spiritual lessons of the nodal axis are going to be a fundamental and very integrated part of that individual's core identity. This also means that balance—between the conscious and the unconscious, the rational and the emotional, the North Node lessons and the South Node gifts—is going to be a very prominent, central theme in that individual's life.

Having any planet conjunct one of the nodes makes it much easier to tap into the energy of that node, because every time you express or experience an activation of the planet, it's also related to the energy of the node. When one is born on a lunar eclipse, it's important to consider which planet is conjunct which node. The South Node, remember, is the easy, familiar territory. It's the place where we bring our skills and experience from the past, but it's also the place where we're most likely to get trapped by old patterns and old habits. The North Node is less familiar—it points the way to how we're meant to use our South Node gifts in this lifetime.

If the Sun is conjunct the South Node and the Moon is conjunct the North Node, the focus will be on learning to be guided by your more intuitive, emotional nature, while still retaining your familiar sense of self and identity shown by the Sun conjunct the South Node. If, on the other hand, the Moon is conjunct the South Node and the Sun is conjunct the North Node, your emotional nature is what is the most comfortable and familiar to you, and your lesson is to learn how to bring that energy into balance and more conscious awareness by working with your Sun conjunct your North Node.

The lessons involved are seen through the signs on the nodal axis, as well as the houses where the nodes (and the Sun and Moon) reside.

One other effect of being born under a lunar eclipse is that you're far more likely to be very sensitive to the transiting lunar eclipses—even when they don't specifically "trigger" anything in your chart. Every lunar eclipse is a recurrence of the lunar eclipse in your natal chart, and every lunar eclipse will bring up issues related to your spiritual path.

Melissa Writes:

"I have a client who has her South Node at 12 degrees Capricorn conjunct her Ascendant at 11 degrees Capricorn. She is considering becoming involved with a man who has Saturn (retrograde) at 11 degrees Capricorn. Any insight? Also, generally what is your insight into the South Node rising so close to the Ascendant?"

Kevin Answers:

Melissa,

There are a number of factors involved here (and aren't there always when the question concerns relationships?). Obviously, I couldn't speculate on whether or not the relationship would be appropriate for your client, but I can share some thoughts on how her nodal axis might come into play.

Whenever we consider the Ascendant, we also have to consider the Descendant. How we approach the world—what we expect when we are expressing ourselves—is always related to what we expect when we're relating to others—when we're receiving input from others. With the nodes so closely conjunct the Ascendant/Descendant axis, your client will naturally encounter her nodal-axis lessons and gifts through relationship both to herself and to others.

The Ascendant/Descendant axis and the 1st house/7th house axis have their own little quirks. What is close to our Ascendant, what is in our 1st house, feels so much a part of ourselves that we have a great deal of difficulty gaining perspective on it. Our Descendant (and our 7th house) is so easily seen that we have difficulty in accepting that it is also a part of us, so we "give away" our 7th house. We project these qualities on to other people in one-to-one relationships until we can finally accept that the people we relate to are actually mirrors, and everything that we see in others is actually a reflection of something that we may have yet to "own" in ourselves.

The Cancer/Capricorn axis has to do with self-responsibility and our ability to get our needs met. The South Node in Capricorn brings the gifts of experience and responsibility, of structure and stability.

These gifts must be balanced through the North Node in Cancer, and we must learn how to open up on an emotional level, to allow ourselves to be nurtured, and to learn how to nurture others. With the South Node in Capricorn rising and the North Node in Cancer setting, your client may tend to take her Capricorn gifts for granted, perceiving them as being a fundamental part of her identity or at least of her approach to the world. On the other hand, she may tend to give away her North Node in Cancer to others, and either feel that she can only have her emotional needs met through a relationship, or that she is incapable of nurturing or being nurtured herself, and must experience this through others. This is not to say that either of these are necessarily the case, but only to suggest that her relationship lessons may involve some of these themes.

By attracting a partner whose Saturn conjoins both her Ascendant and her South Node, she is attracting someone who will very probably force her to take a much greater amount of self-responsibility. She could experience this as being extremely supportive, or as extremely confining and repressive, depending entirely on where she is with her personal growth, and on how well she has learned to work with her own Saturn.

Michelle Writes:

"I was wondering about the effects of an eclipse. I understand that a solar eclipse will affect you for one year and a lunar eclipse for up to six months. What happens if the eclipse occurs conjunct one of your natal planets? I have the solar eclipse coming up conjunct my Venus in the 7th house. I'm confused about how these energies play out. What about a lunar eclipse opposite Pluto? Is this a sure sign of trouble to come?"

Kevin Answers:

Michelle,

I'm going to start out with a little background information on eclipses, and then talk a bit about how we can interpret them.

Eclipses are very important events astrologically, and yet they are also very misunderstood (much like the Moon's nodes, which just happen to be the transiting eclipse points). Most people think that eclipses are rare events, but actually we have two sets of eclipses every year, or one set of eclipses every five-and-a-half months or so. What is rare is an eclipse that can actually be observed. Solar eclipses can only be seen across a relatively narrow path. Lunar eclipses, on the other hand, can be seen from just about every place that happens to be experiencing night at the time of the eclipse.

Solar eclipses can only occur at the moment of the New Moon; lunar eclipses can only occur at the moment of the Full Moon. You can see eclipses in a chart by simply looking at the Moon's nodes. Whenever there is a Full Moon or a New Moon within 18°31' of one of the nodes, it is an eclipse. The Moon's nodes represent the points where the orbit of the Moon crosses the ecliptic (the apparent orbit of the Sun around the Earth, but actually the orbital plane of the Earth around the Sun). Whenever there is a lunation (New or Full Moon) within 18°31' of the nodes, the Sun, Moon, and Earth are not only lining up by longitude (sign), but also by declination, and the Moon's shadow will at least partially block the Sun (in a solar eclipse) or the

Earth's shadow will block the Sun's light from reaching the Moon (in a lunar eclipse).

Let's talk about interpreting eclipses in general first, and then move on to the specifics. Eclipses are the single most significant astrological events, and their effects can be felt for up to six months at a time (or from one eclipse cycle until the next eclipse cycle). Even when an eclipse does not contact personal planets in our chart, eclipses are always significant. Astrologer Robert Jansky says that eclipses represent areas of crisis; the houses in which the eclipses fall in our charts represent areas of life that will receive the most focus over the next few months, and that will frequently undergo some sort of a change.

Eclipses usually occur in pairs, although it is possible to have three eclipses in a cycle (either solar-lunar-solar, or lunar-solar-lunar). The degrees of the eclipses will be sensitive degrees for the next six months or so, both on a personal level and on a mundane level.

In a solar eclipse, the Moon passes between the Sun and the Earth, temporarily blocking out the light of the Sun. On an interpretive level, a solar eclipse represents the past, our unconscious nature (the Moon) overtaking and overshadowing our conscious expression of self (the Sun). More than this, what we "see" is the dark, hidden side of the Moon. In a solar eclipse we are forced to look at our shadow self, to acknowledge it and recognize it or else it will consume us.

Solar eclipses tend to relate to sudden events that disrupt our normal conscious functioning; they are the crises that seem to come out of nowhere and that suddenly demand all of our conscious attention and focus. We must recognize that the situations that arise as a result of a solar eclipse operate on two levels: Once we've addressed the manifest crisis, it's important to remember that no matter how much it seems like the crisis is coming from the outside world, it's actually a projection from our own unconscious and subconscious. By acknowledging and accepting these parts of ourselves, we can help keep them from disrupting our lives periodically. (The important lesson is that we must accept and acknowledge our shadow self rather than deny and ignore it.)

In a lunar eclipse, on the other hand, the Earth moves between the Sun and the Moon at the Full Moon, casting a shadow on the Moon at the moment of the Full Moon and blocking the Sun's light from reflecting on the Moon. Lunar eclipses are significant disruptions of the lunar cycle. Normally, what we have started during the New Moon will grow and change until the Full Moon, when it reaches fruition and is illuminated and we are able to understand and work with it on a conscious level; after the Full Moon, the light is distributed and we make use of our new knowledge and prepare for the next cycle.

During a lunar eclipse, however, that moment of illumination is suddenly gone. In many ways, it's like having the rug suddenly pulled out from under your feet. Individuals who are very sensitive to the lunar cycles often find lunar eclipses to be very disorienting. My personal sense of lunar eclipses is that they represent opportunities to break out of our current cycle and move on to a different level. The window of opportunity here is rather brief, and we have to act on it consciously; but lunar eclipses provide an opportunity to make a kind of evolutionary leap, to demonstrate a certain degree of insight and proficiency in whatever our current lesson may be, and to be allowed to move on to the next phase without having to take the long way around. Please remember that we are not in a race and that taking advantage of these opportunities does not make one a better person or "more spiritual" or "more highly evolved."

In any event, a lunar eclipse can make us more aware of our cycles and patterns, as well as of our unconscious motivations. When the cycle isn't allowed to complete, when the Earth's shadow blocks out the light of the Sun, we experience the same type of disorientation as when we're doing something routine and not thinking about it and are suddenly interrupted in the middle of it. We can either catch ourselves and decide to take a different course of action, or else we can try to continue what it was that we were doing; but if we try to complete the cycle, it is with a new awareness: suddenly we have to think about what we're doing a bit more.

Now, what happens when an eclipse makes an aspect to a personal planet? First of all, the general rule that eclipses represent crisis, focus, and change applies. If an eclipse makes a close aspect to a planet in your chart, that planet is going to be involved in the changes and activities of the eclipses. In particular, if an eclipse is conjunct a planet in your chart, then the function of that planet will be tied in with your personal experience of the eclipse.

Again, solar eclipses tend to stir things up on a physical and conscious level much more than lunar eclipses do (although the lunar eclipse can make us acutely aware of our emotional cycles!). A solar eclipse conjoining your Venus, for example, might indicate that your unconscious relationship issues and needs will be coming to the forefront in the next few months, and will require some conscious and immediate attention. A lunar eclipse conjunct Venus might have more to do with relationship patterns and cycles.

The last thing to remember about eclipses is that their effects do last for up to six months. The key events may occur within the first month and our lives may get stirred up quite a bit, but we do have the next four or five months to understand and integrate the lessons.

Thomas Writes:

"How will the total lunar eclipse affect us? What is the difference between a total and partial eclipse astrologically? We do not see them very often—does this have a special effect?"

Zodi Writes:

"What is the best way to 'use' a lunar eclipse in a Solar Return chart (and how rare is this occurrence)? The Moon is also the most aspected planet in the chart."

Kevin Answers:

Thomas and Zodi,

Lunar eclipses aren't rare—we get at least two lunar eclipses every year. Lunar eclipses are also much easier to observe than solar eclipses are. A solar eclipse is only visible in a very small, specific area, while a lunar eclipse is visible from anywhere on Earth where it happens to be night. While there's certainly a difference astronomically between a total and a partial lunar eclipse, there's no difference astrologically between them.

Lunar eclipses tend to be far less important than solar eclipses. Lunar eclipses that closely aspect planets in the natal chart may prompt emotional and psychological changes, but the big changes are more closely tied to the solar eclipses. The main reason for this is that while lunar eclipses occur at a Full Moon (the halfway point of a cycle), solar eclipses are at the New Moon (the beginning of a new cycle). Once a cycle is underway, all we can do is ride it out; we can't really instigate any significant changes until the old cycle ends and the new one begins. This is why lunar eclipses tend to highlight what's already there, rather than bring entirely new challenges and lessons to the picture.

As far as the lunar eclipse in the Solar Return chart goes, it really depends on the house placement of the Sun and Moon, and the planets aspecting the Moon. In a Solar Return chart, the *only* houses to look at are the angular houses (1, 4, 7, 10). Any planet not in an angular house is not going to have any impact, no matter what it aspects. The Sun and Moon are the only exceptions to this rule–they're always important. They're the strongest in angular houses, and the weakest in cadent houses (3, 6, 9, 12). A weak Sun and Moon, particularly a 6th-house/12th-house emphasis, can sometimes be an indication of illness. The only aspects to the Sun and Moon that you need to pay attention to are aspects made from planets in angular houses–these are going to be very prominent ("foreground") energies for the period of the return. While having a lunar eclipse in a Solar Return chart is somewhat unusual, it's bound to happen eventually (especially if you're working with quarti- and demi-returns). In and of itself, I wouldn't consider it to be a terribly important factor.

Astrology
Understanding the Birth Chart

KEVIN BURK

This beginning-to intermediate-level astrology book is based on a course taught to prepare students for the NCGR Level I Astrological Certification exam. It is a unique book for several reasons. First, rather than being an astrological phrase book or "cookbook," it helps students to understand the language of astrology. From the beginning, students are encouraged to focus on the concepts, not the keywords. Second, as soon as you are familiar with the fundamental elements of astrology, the focus shifts to learning how to work with these basics to form a coherent, synthesized interpretation of a birth chart. In addition, it explains how to work with traditional astrological techniques, most notably the essential dignities. All interpretive factors are brought together in the context of a full interpretation of the charts of Sylvester Stallone, Meryl Streep, Eva Peron, and Woody Allen. This book fits the niche between cookbook astrology books and more technical manuals.

- Discover how classical astrology can enrich your understanding of the planets, signs, and houses
- Use the essential dignities to determine the relative strength or weakness of a planet in a particular sign
- Explore the methodology behind the different systems of house division
- Discover the mechanics and the effects of the Moon's nodes
- Study aspect patterns and their effects in the chart
- Use the comprehensive worksheet to lead you through all the interpretive factors necessary

1-56718-088-4, 384 pp., 7½ x 9⅛, illus. $17.95

To order, call 1-877-NEW-WRLD
Prices subject to change without notice

The New A to Z Horoscope Maker and Delineator
Llewellyn George

A textbook . . . encyclopedia . . . self-study course . . . and extensive astrological dictionary all in one! More American astrologers have learned their craft from *The New A to Z Horoscope and Delineator* than any other astrology book.

First published in 1910, it is in every sense a complete course in astrology, giving beginners all the basic techniques and concepts they need to get off on the right foot. Plus it offers the more advanced astrologer an excellent dictionary and reference work for calculating and analyzing transits, progression, rectifications, and creating locality charts. This new edition has been revised to met the needs of the modern audience.

0-87542-264-0, 592 pp., 6 x 9   **$17.95**

Composite Charts

The Astrology of Relationships

JOHN TOWNLEY

How does the world see you as a couple? Will your battles bring your closer together or waste your time? What is the best way to keep the romance alive? The composite chart describes that special dynamic that makes a couple more than the sum of its two personalities. It is a new, mathematically produced horoscope made up of mutual midpoints between the natal charts of two individuals.

Composite Charts is the definitive work on relationship dynamics by the "father of the composite chart," who has spent more than twenty years developing the technique he introduced in 1973. It incorporates a systematic theory of how and why composite charts work, comprehensive sections on composite planets in signs, houses, and aspects, as well as on the interplay of the natal charts with the composite chart they form, revealing who runs what within the relationship—or whether the relationship itself runs its progenitors.

1-56718-716-1, 528 pp., 7½ x 9⅛   **$24.95**

Heaven Knows What
GRANT LEWI

Here's the fun, new edition of the classic *Heaven Knows What*! What better way to begin the study of astrology than to actually do it while you learn. *Heaven Knows What* contains everything you need to cast and interpret complete natal charts without memorizing any symbols, without confusing calculations, and without previous experience or training. The tear-out horoscope blanks and special "aspect wheel" make it amazingly easy.

The author explains the influence of every natal Sun and Moon combination, and describes the effects of every major planetary aspect in language designed for the modern reader. His readable and witty interpretations are so relevant that even long- practicing astrologers gain new psychological insight into the characteristics of the signs and meanings of the aspects.

Grant Lewi is sometimes called the father of "do-it-yourself" astrology, and is considered by many to have been astrology's forerunner to the computer.

0-87542-444-9, 480 pp., 6 x 9, tables, charts   **$14.95**